The Handbook
of Environmental Chemistry

Volume 3   Part G

Edited by O. Hutzinger

Anthropogenic Compounds

With contributions by
F. Brochhagen, E. P. Burrows, H. Fiedler,
J. Konietzko, W. R. Mitchell, K. Mross, W. Mücke,
D. L. Parmer, D. H. Rosenblatt

With 25 Figures and 75 Tables

Springer-Verlag
Berlin Heidelberg GmbH

Professor Dr. Otto Hutzinger

University of Bayreuth
Chair of Ecological Chemistry and Geochemistry
Postfach 101251, W 8580 Bayreuth, FRG

ISBN 978-3-662-15019-1

Library of Congress Cataloging-in-Publication Data.

(Revised for Volume 3/G) Anthropogenic compounds. (The Handbook of environmental chemistry;
v. 3) Spine title, pt. E: Compounds. Includes bibliographical references and indexes.
1. Pollution – Environmental aspects. 2. Pollutants – Toxicology. 3. Environmental chemistry.
I. Anliker, R. (Rudolf), 1926 –. II. Title: Compounds. III. Series.
QD31.H335 vol. 3 [QH545.A1] 574.5'222 s 80–16609
ISBN 978-3-662-15019-1   ISBN 978-3-540-46757-1 (eBook)
DOI 10.1007/978-3-540-46757-1

© Springer-Verlag Berlin Heidelberg 1991
Originally published by Springer-Verlag Berlin Heidelberg New York in 1991
Softcover reprint of the hardcover 1st edition 1991

Typesetting: Macmillan India Ltd., Bangalore-25

2152/3020-543210 – Printed on acid-free paper

Preface

Environmental Chemistry is a relatively young science. Interest in this subject, however, is growing very rapidly and, although no agreement has been reached as yet about the exact content and limits of this interdisciplinary subject, there appears to be increasing interest in seeing environmental topics which are based on chemistry embodied in this subject. One of the first objectives of Environmental Chemistry must be the study of the environment and of natural chemical processes which occur in the environment. A major purpose of this series on Environmental Chemistry, therefore, is to present a reasonably uniform view of various aspects of the chemistry of the environment and chemical reactions occurring in the environment.

The industrial activities of man have given a new dimension to Environmental Chemistry. We have now synthesized and described over five million chemical compounds and chemical industry produces about one hundred and fifty million tons of synthetic chemicals annually. We ship billions of tons of oil per year and through mining operations and other geophysical modifications, large quantities of inorganic and organic materials are released from their natural deposits. Cities and metropolitan areas of up to 15 million inhabitants produce large quantities of waste in relatively small and confined areas. Much of the chemical products and waste products of modern society are released into the environment either during production, storage, transport, use or ultimate disposal. These released materials participate in natural cycles and reactions and frequently lead to interference and disturbance of natural systems.

Environmental Chemistry is concerned with *reactions in the environment*. It is about distribution and equilibria between environmental compartments. It is about reactions, pathways, thermodynamics and kinetics. An important purpose of this Handbook is to aid understanding of the basic distribution and chemical reaction processes which occur in the environment.

Laws regulating toxic substances in various countries are designed to assess and control risk of chemicals to man and his environment. Science can contribute in two areas to this assessment: firstly in the area of toxicology and secondly in the area of chemical exposure. The available concentration ("environmental exposure concentration") depends on the fate of chemical compounds in the environment and thus their distribution and reaction behaviour in the environment. One very important contribution of Environmental Chemistry to the above mentioned toxic substances laws is to develop laboratory test methods, or mathematical correlations and models that predict the environmental fate of new chemical compounds. The third purpose of this Handbook is to help in the basic understanding and development of such test methods and models.

The last explicit purpose of the handbook is to present, in a concise form, the most important properties relating to environmental chemistry and hazard assessment for the most important series of chemical compounds.

At the moment three volumes of the Handbook are planned. Volume 1 deals with the natural environment and the biogeochemical cycles therein, including some background information such as energetics and ecology. Volume 2 is concerned with reactions and processes in the environment and deals with physical factors such as transport and adsorption, and chemical, photochemical and biochemical reactions in the environment, as well as some aspects of pharmacokinetics and metabolism within organisms. Volume 3 deals with anthropogenic compounds, their chemical backgrounds, production methods and information about their use, their environmental behaviour, analytical methodology and some important aspects of their toxic effects. The material for volumes 1, 2, and 3 was more than could easily be fitted into a single volume, and for this reason, as well as for the purpose of rapid publication of available manuscripts, all three volumes are published as a volume series (e.g. Vol. 1; A, B, C). Publisher and editor hope to keep the material of the volumes 1 to 3 up to date and to extend coverage in the subject areas by publishing further parts in the future. Readers are encouraged to offer suggestions and advice as to future editions of "The Handbook of Experimental Chemistry".

Most chapters in the Handbook are written to a fairly advanced level and should be of interest to the graduate student and practising scientist. I also hope that the subject matter treated will be of interest to people outside chemistry and to scientists in industry as well as government and regulatory bodies. It would be very satisfying for me to see the books used as a basis for developing graduate courses on Environmental Chemistry.

Due to the breadth of the subject matter, it was not easy to edit this Handbook. Specialists had to be found in quite different areas of science who were willing to contribute a chapter within the prescribed schedule. It is with great satisfaction that I thank all authors for their understanding and for devoting their time to this effort. Special thanks are due to the Springer publishing house and finally I would like to thank my family, students and colleagues for being so patient with me during several critical phases of preparation for the Handbook, and also to some colleagues and the secretaries for their technical help.

I consider it a privilege to see my chosen subject grow. My interest in Environmental Chemistry dates back to my early college days in Vienna. I received significant impulses during my postdoctoral period at the University of California and my interest slowly developed during my time with the National Research Council of Canada, before I was able to devote my full time to

Environmental Chemistry in Amsterdam. I hope this Handbook will help deepen the interest of other scientists in this subject.

Otto Hutzinger

This preface was written in 1980. Since then publisher and editor have agreed to expand the Handbook by two new open-ended volume series: Air Pollution and Water Pollution. These broad topics could not be fitted easily into the headings of the first three volumes.

All five volume series will be integrated through the choice of topics covered and by a system of cross referencing.

The outline of the Handbook is thus as follows:

1. The Natural Environment and the Biogeochemical Cycles
2. Reactions and Processes
3. Anthropogenic Compounds
4. Air Pollution
5. Water Pollution

Bayreuth, January, 1991   Otto Hutzinger

Contents

List of Contributors

Dr. Franzkarl Brochhagen
Max-Bruch-Str. 3
W5060 Bergisch-Gladbach 2
FRG

Elizabeth Parker Burrows
3316 Old Forest Road
Baltimore, MD 21208
USA

Dr. Heidelore Fiedler
Lehrstuhl für Ökologische
Chemie und Geochemie
Universität Bayreuth
Postfach 10 12 51
W 8580 Bayreuth
FRG

Prof. Dr. J. Konietzko
Institut für Arbeits-
und Sozialmedizin
Obere Zahlbacher Str. 67
Johannes-Gutenberg-Universität
W 6500 Mainz
FRG

Wayne R. Mitchell
3316 Old Forest Road
Baltimore, MD 21208
USA

Dr. K. Mross
Institut für Arbeits-
und Sozialmedizin
Obere Zahlbacher Str. 67
Johannes-Gutenberg-Universität
W 6500 Mainz
FRG

Dr. W. Mücke
Lehrstuhl für Ökologische
Chemie und Geochemie
Universität Bayreuth
Postfach 10 12 51
W 8580 Bayreuth
FRG

David L. Parmer
3316 Old Forest Road
Baltimore, MD 21208
USA

Dr. David H. Rosenblatt
3316 Old Forest Road
Baltimore, MD 21208
USA

Isocyanates

Franzkarl Brochhagen

Max-Bruch-Str. 3, D-5060 Bergisch Gladbach 2

The Handbook of Environmental Chemistry,
Volume 3 Part G, Ed. O. Hutzinger
© Springer-Verlag Berlin Heidelberg 1991

Summary

The worldwide consumption of isocyanates (1985) is in the order of 1.5 million metric tons per annum. Monoisocyanates which represent less than 3 percent of the total tonnage are used as intermediates for the synthesis of pesticides. Di- and polyisocyanates are the key raw materials for a class of polymers which are known under the collective name polyurethanes and which are manufactured by numerous companies for a great variety of uses.

    As isocyanates are potentially hazardous materials particularly if their vapor is inhaled or if they come into contact with the skin it is essential to have full information available on the health risk which may occur due to improper handling and on a possible impact on the environment.

    The literature relating to the toxicological, ecological, analytical and occupational health aspects of these chemicals is evaluated, the international regulatory situation is presented. Recommendations on safe handling of isocyanates, on workplace monitoring and on occupational hygiene measures are given. The efforts of the isocyanate manufacturers to generate additional information in this area by jointly sponsoring research projects are illustrated.

1 Introduction – Historical

It is almost years ago – namely in 1849 – since Wurtz synthesized the first organic isocyanate [1]. The marked activity of this new class of products and the specific molecular structure of the isocyanate group stimulated the interest of researchers during the following decades. These efforts resulted in various laboratory scale methods of synthesis and subsequently in numerous new compounds bearing one or two isocyanate groups in the molecule.

It was not however until one century after the basic discovery that industrial production and use of these chemicals for the manufacture of a new class of polymers were considered. In 1947 this new orientation was characterized by a British author as follows: "Many of organic chemicals discovered and studied in the nineteenth century heyday of synthetic organic chemistry and subsequently consigned to the dust-laden leaves of > Annalen < and > Berichte < have been brought back into the light of day by the perspicacity of the plastics research workers. As a result of vigorous development it is possible today to cite several commercial chemicals in quantiy production as plastics raw materials which only a few years back were regarded as expensive laboratory curiosities. To some extent this is true of the isocyanates" [2].

O. Bayer and his co-workers in 1937 applied for a patent on the "Diisocyanate-Polyaddition-Process"i.e. the controlled synthesis of macromolecules whose properties can be tailor-made for manifold applications [3,4]. The commercialization of these polymers, for which the collective name "Polyurethanes" has become customary, started in the early fifties. Production capacities for chemicals as well as know-how and processing techniques were provided by the chemical industry and engineering firms in order to serve companies which undertook to manufacture solid or cellular polyurethane plastics as semi-finished materials or finished articles for a great variety of uses. Coatings, adhesives, textile and leather finishes as well as specific fibers were also available by this new process.

In 1985, the worldwide consumption of di- and polyisocyanates for the manufacture of polyurethanes was 1.5 million metric tonnes in total [5]. Toluene diisocyanate (TDI) and diphenylmethane diisocyanate (MDI) represent more than 95% of this tonnage. They are marketed as commodity chemicals. Hexamethylene-(HDI), isophorone-(IPDI) and 1,5-naphthylene diisocyanate (NDI) are used to a considerable extent for specialty polyurethanes in the field of elastomers and coatings.

In contrast, the world consumption (1985) of monoisocyanates used as intermediates for the production of pesticides and insecticides is in the order of 40000 metric tonnes per annum [6].

Of the various monoisocyanates used on an industrial scale, the methyl- and 3,4-dichlorophenyl compounds are the most important.

Isocyanates are potentially hazardous materials, in particular if their vapor is inhaled. It is therefore evident that the degree of hazard depends on the volatility of these chemicals which are mostly liquid at ambient temperature.

Researchers in the early stage describe monoisocyanates as strong lachrimators, extremely irritating to the eyes and the respiratory tract. The occurrence of asthmatic symptoms was observed and related to inhalation. Skin contact was also found to cause irritation often followed by a "tanning" effect.

To date, there is comprehensive medical knowledge available on the health risk which may result from exposure to isocyanates. Occupational hygiene measures and "Threshold Limit Values" for the admissible concentration in the workplace have been established ensuring maximum protection for employees handling these chemicals.

Isocyanates react readily with water. The reaction products of diisocyanates are carbon dioxide and mainly polyurea which is practically inert and hardly biodegradable. For this reason, there is no specific problem with regard to the pollution of air, water and soil as diisocyanates are converted to mostly harmless materials rather quickly [see 2.2.2].

Only limited information is available on monoisocyanate reaction products formed after reaction with water. There is no doubt that ureas are generated as well.

By establishing the I.I.I.–International Isocyanate Institute Inc.–[7] in 1972 the isocyanate producers in the Far East, The Americas and Western Europe started a joint effort to pool environment related information they have available, especially on TDI and MDI. The main objective of the I.I.I is to sponsor and to finance research projects related to toxicological, occupational health, analytical and environmental aspects of these two most important isocyanates. Various projects have been finalized in the meantime and are the subject of respective publications.

2 General

Isocyanates can be described by the general formula R–N=C=O in which R is an aliphatic, araliphatic, alicyclic, aromatic or heterocyclic substituent. According to the number of –N=C=O groups in the molecule one has to distinguish between mono-, di- and polyisocyanates, (see also [125] definition of polyfunctional IC). The chemistry of the –N=C=O group is linked to the specific electrophilic character of the C-atom which can be expressed by the two resonance structures

$$R-\bar{N}-\overset{+}{C}=O \longleftrightarrow R-N=C=O \longleftrightarrow R-N=\overset{+}{C}-\bar{O}$$

These structures explain why isocyanates undergo numerous addition reactions and participate in the formation of cyclic compounds.

Production methods for isocyanates and their reaction behaviour are the subject of various monographs and reviews [4,8–15].

In the following sections laboratory synthesis procedures are only mentioned in brief; industrial production techniques are dealt with in more detail with particular emphasis on future possibilities. Specific attention is focused upon reactions which are of importance with regard to the industrial use of isocyanates, their environmental fate and their possible interactions with biological materials.

2.1 Production Methods

2.1.1 Alkyl- and Arylisocyanates

The most important and elegant method of manufacturing isocyanates is via the reaction of primary amines or their salts with phosgene:

$$R-NH_2 + COCL_2 \longrightarrow R-N=C=O + 2\,HCl \qquad (1)$$

$$R-NH_2 \cdot HCl + COCl_2 \longrightarrow R-N=C=O + 3\,HCl \qquad (2)$$

A synthesis on the basis of reaction (2) was described in 1884 [16]; an isocyanate was obtained by combining phosgene with the salt of a primary amine. Methylisocyanate was first synthesized by treating molten methylamine hydrochloride with phosgene [17]. The resulting methylcarbamoylchloride was decomposed upon heating with lime:

$$CH_3-NH_2 \cdot HCl + COCl_2 \longrightarrow CH_3-NH-COCl + 2\,HCl \qquad (3)$$

$$2\,CH_3-NH-COCl + 2\,CaO \longrightarrow$$
$$2\,CH_3-NCO + CaCl_2 + Ca(OH)_2 \qquad (4)$$

With aliphatic diamines it has proven preferable to use the salt which is obtained by reaction with carbon dioxide for the phosgenation process [13]. The treatment of ureas and of carbamic acid esters with phosgene also offers a possibility for the synthesis of isocyanates [18,19].

   The direct phosgenation of primary amines has by far the greatest importance for industrial production. The process is rather complex; the reaction conditions, e.g. the quantity relation of the reaction partners, the temperature and the residence time have a significant influence on the yield of the final product. The following equations illustrate the occurence of intermediates and possible side reactions in this process [12,20]:

$$R-NH_2 + COCl_2 \longrightarrow R-NH-COCl + HCl \qquad (5)$$

$$R-NH_2 + HCl \longrightarrow R-NH_2 \cdot HCl \qquad (6)$$

$$R-NH_2 + R-NH-COCl \longrightarrow R-N=C=O + R-NH_2 \cdot HCl \qquad (7)$$

$$R-NH_2 + R-N=C=O \longrightarrow R-NH-CO-NH-R \qquad (8)$$

$$R-NH-COCl \rightleftharpoons R-N=C=O + HCl \qquad (9)$$

$$R-NH-CO-NH-R + COCl_2 \longrightarrow 2\,R-NH-COCl \qquad (10)$$

Furthermore, the formation of biurets, carbodiimides, uretdions and isocyanurates is also possible. In order to avoid the formation of (poly)ureas, excess phosgene is normally used (up to 200% of the chemical equivalent). This is because the formation of isocyanates from ureas and phosgene is not quantitative and unfavorably influences the reaction time. The primary reaction leading to the carbamoylchloride is exothermic, whereas the process as a whole is endothermic.

The phosgenation process is usually performed in an inert solvent with a boiling point lower than that of the isocyanate. Chlorobenzene and o-dichlorobenzene are used in the production of TDI and MDI. For the production of lower boiling point monoisocyanates, phosgenation in the gas phase at elevated temperatures is the method employed. Vapors of the primary amine and phosgene are combined in a reaction tube and react with each other within a very short residence time. Catalysis is not necessary for the vapor phase reaction nor for the solvent process. For details on the layout of production units and a survey of the extensive patent literature, see [12,13].

In view of the toxicity of phosgene, numerous efforts have been made in the past to develop "phosgene-free" technologies for the production of isocyanates on an industrial scale. From the various laboratory synthetic methods available [13] such as

1. thermolytic decomposition of carbamic acid esters
2. thermolysis of ureas
3. alkylation of alkali cyanates
4. reaction of isocyanic acid with olefins
5. rearrangement of N-containing derivatives of carboxylic acids

2. has become the basis of a production process for methylisocyanate [21]. N,N'-dimethyl urea is combined with diphenylester of carboxylic acid at temperatures of 200 . . . 250° according to the following equation:

$$CH_3-NH-CO-NH-CH_3 + C_6H_5-O-CO-O-C_6H_5$$
$$\longrightarrow 2 \ R-N=C=O + 2 \ C_6H_5-OH \tag{11}$$

Ethyl, propyl- and butylisocyanate can also be obtained by this process.

A technology which has been studied thoroughly in the recent past is the reaction of carbon monoxide with aromatic nitro-, nitroso-, azo-, azoxy- and acido-compounds [13]. The following summary equation illustrates this reaction for dinitro toluene:

$$CH_3-C_6H_3-(NO_2)_2 + 6 \ CO$$
$$\longrightarrow CH_3-C_6H_3-(N=C=O)_2 + 4 \ CO_2 \tag{12}$$

This process requires rather elevated pressures and high temperatures and the presence of relatively high concentrations of catalyst containing rhodium or palladium.

As the production of isocyanates by carbonylation of aromatic nitro compounds is possible in a one synthesis step and the conventional phosgenation is a two step procedure, it was expected that the direct synthesis would offer commercial advantages assuming good yields and a limited loss of catalyst. In spite of intensive research efforts specifically aimed at the catalytic problem, this method has not been successful for the industrial production of isocyanates.

The carbonylation of aromatic nitro compounds in the presence of alcohols leads to the formation of N-aryl carbamic acid alkylesters which are obtained in very good yield. Considerably lower concentrations of precious metal catalysts are needed than is the case for direct carbonylation to isocyanates. Moreover, selenium or sulfur can be used as catalysts. Isocyanates are generated by thermal decomposition of these carbamic acid esters. There is likewise no industrial application of this process yet.

2.1.2 Isocyanates with Special Substituents

Numerous isocyanates with various substituents have been synthesized in the past [9,13]. Details are not given here as these products are only of interest for research purposes.

2.1.3 Analysis and Quality Control

In application, isocyanates with a high degree of purity are required. This is essential for monoisocyanates which are used as intermediates and particularly true for diisocyanates applied in the manufacture of polymers. The primary characteristic is the $N=C=O$–content which can be determined by adding dibutylamine in excess to a weighed sample of the product. After reaction with the isocyanate is complete, the $N=C=O$–content is obtained by titration determining the unreacted amine.

The chlorine (total and hydrolysable) and iron content are important quality features as they can increase or decrease the reactivity of the $-N=C=O-$ group. Control of physical parameters such as refraction index and viscosity can contribute to assure a constant quality. For detailed analytical methods see [22, 23, 24].

2.2 Chemistry and Typical Reactions

The chemistry of isocyanates is extremely complex. The strongly electrophilic character of the C-atom in the $N=C=O$-group is responsible for their ability to react with compounds containing an "active" hydrogen atom:

$$R-N=C=O + H-X \longrightarrow R-NH-CO-X \tag{13}$$

This "addition" usually takes place at ambient or moderately elevated temperatures. Depending on the organic substituent, the rate of reaction can vary: aliphatic isocyanates are normally less reactive than aromatic ones; the two N=C=O-groups in diisocyanates have different activities because of the activating effect of one N=C=O-group towards the other. Catalysts such as *t*-aliphatic amines and organo-metallic compounds accelerate the rate of reaction.

In most cases, the addition product is thermally stable up to 150 . . . 180°. As a rule, it dissociates at higher temperatures forming the initial reactants again. The hydrogen in the NH-group of urethane or urea bonds can further react with –N=C=O, thus forming allophanates or biurets.

Also possible is the dimerization and trimerisation of isocyanates leading to uretdions and isocyanurates, respectively. A number of specific phosphorous compounds, amines and alkali salts of organic acids can be used as catalysts. Condensation of two N=C=O– groups produces carbodiimides; carbon dioxide is split off in this reaction for which some specific phosphorous compounds are suitable catalysts:

$$2 \text{ R-N=C=O} \longrightarrow \text{R-N=C=N-R} + CO_2 \tag{14}$$

Several notable surveys of isocyanate reactions and reaction kinetics have appeared in the past [11, 13, 25–28]. Some typical reactions of isocyanates which are important for their use as chemical intermediates and as raw materials for polyurethanes as well as their interaction with substances in the environment and biological substrates require special attention.

2.2.1 Alcohols and Amines

The result of the addition of isocyanates to hydroxyl group bearing molecules is an ester of carbamic acid which is characterized by the *urethane* group:

$$\text{R}'\text{-N=C=O} + \text{R}''\text{-OH} \longrightarrow \text{R}'\text{-}NH\text{-}CO\text{-}O\text{-}R'' \tag{15}$$

If a diol is combined with double the stoichiometric amount of a diisocyanate, a "prepolymer" with two urethane and two –N=C=O groups is formed:

$$2 \text{ R-N=C=O} + \text{HO-} (CH_2)_4\text{-OH} \longrightarrow$$

$$\text{O=C=N-R-NH-CO-O-} (CH_2)_4\text{-O-CO-NH-R-N=C=O} \tag{16}$$

It is evident that the combination of a diol with an equimolecular amount of isocyanate results in a linear *polyurethane* polymer.

Polyester or polyether polyols with hydroxyl end groups are mainly used as reaction partners of diisocyanates for the manufacture of polyurethane plastics. The degree of branching and the resulting functionality of the polyol resin determines the extent of cross-linking in the final product. Flexible, semi-rigid or rigid polyurethanes can be obtained by varying these parameters; the hardness of the final product can also be increased by raising the diisocyanate above the equimolar amount. The excess then reacts with the hydrogen atoms of the urethane NH-groups and other active sites.

The reaction of the $-N=C=O$ group with primary amines to substituted *ureas* is characterized by the following equation:

$$R-N=C=O + R''NH_2 \longrightarrow R-NH-CO-NH-R'' \tag{17}$$

If the hydrogen of the urea reacts further with isocyanates *biuret* structures are formed. Primary aliphatic or aromatic amines are often used as cross linkers for polyurethanes. They react with the respective equivalent of additional diisocyanate.

2.2.2 Water

The solubility of monoisocyanates in water is rather low but gradually differs and depends on the type of substituent (see 3.3.1). Diisocyanates however exhibit only an extremely low degree of solubility. The dissolved portion reacts rather quickly forming the instable *carbamic acid* which is decomposed to a primary amine and carbon dioxide:

$$R-N=C=O + H_2O \longrightarrow$$
$$R-NH-COOH \longrightarrow R-NH_2 + CO_2 \tag{18}$$

The amine reacts with free isocyanate (if available) to *urea*:

$$R-NH_2 + R-N=C=O \longrightarrow R-NH-CO-NH-R \tag{19}$$

The carbamic acid can alternatively react further with free isocyanate (if available) forming the anhydride of carbamic acid:

$$R-NH-COOH + R-N=C=O \longrightarrow$$
$$R-NH-CO-O-CO-NH-R \tag{20}$$

The anhydride looses carbon dioxide and urea is also formed.

It is obvious that diisocyanates are converted to polyureas. These are solid substances which are hardly fusible and are insoluble in water [11].

The formation of carbon dioxide from water and isocyanate is important for the manufacture of polyurethane foams. Well defined amounts of water are therefore often added to the mixture of polyol and diisocyanate. The carbon dioxide assures the expansion of the reacting components; its amount controls the density of the final product. The polyurea becomes part of the polymer network.

It must be noted that the result of studies into the behavior of living organisms opposite isocyanates in an aquatic environment can only be interpreted with difficulty as it is not fully clear what effects, if any, are related to the isocyanate and/or to its hydrolysis products.

2.2.3 Biological Substrates

If isocyanates are inhaled or administered otherwise, a reaction with water must be expected. Due to their high reactivity an interaction with proteins may take place as well. Various studies into this problem have been conducted, most of them since the early seventies which were the subject of a literature evaluation [29]. The main conclusions were:

1. Basic in vitro studies on the reaction of monoisocyanates with amino acids, peptides and proteins have shown that free NH_2- and OH-groups are the preferred reaction sites. Diisocyanates have been studied as coupling agents to combine a protein with a "label" compound. This is a model reaction for the in vitro synthesis of antigens on the basis of human serum albumin. The possible influence of by-products formed in the aqueous environment has been generally disregarded in these studies [30–39].
2. Monoisocyanates influence the activity of enzymes: in most cases inactivation occurs. The protein SH-group is the preferred binding site. The in vitro cholinesterase inhibition by various isocyanates has pharmacological implications. Side reactions resulting from the aqueous environment were to some extent taken into account [40–60].
3. Tolyl mono- and diisocyanate, when reacted in vitro with human serum albumin, form a synthetic antigen with a tolyl hapten group. This finding contributed to clarifying a postulate according to which the formation of an antigen in vivo would prompt the formation of *IgE* antibodies. It is confirmed to some extent that there is a relationship between the occurrence of such antibodies and symptoms of isocyanate hypersensitivity [61, 62, 63]. A number of more recent publications deal with this aspect (see 5.1.5).

3 Monoisocyanates

As has already been mentioned, monoisocyanates represent less than 3 percent of the total world consumption of isocyanates. They nevertheless play an important part as intermediates in the production of pesticides and of some pharmaceutical specialities. Numerous products have been studied with regard to their suitability for this area of application. Only a limited number have achieved appreciable importance in the industry which are the subject of discussion in the following sections.

3.1 Products, Capacities and Consumption

The monoisocyanates included in two market surveys [6, 64] are shown in Table 1 ("isocyanate" is abbreviated "IC" in the tables and the following text in as far as they relate to monoisocyanates). An estimate of the world capacity and consumption is presented in Table 2. The figures indicate that the consumption is in the order of approximately 55 percent of the capacity. Tables 3 and 4 show the consumption of various mono-ICs in the United States and in Western Europe. It is obvious that methyl-IC is the predominant product in the USA whereas 3,4-dichlorophenyl-IC has by far the greatest importance in Europe. According to [6] methyl-IC is manufactured in Israel and Japan, exclusively. The mono-IC producing companies are listed in Table 5.

Table 1. Monoisocyanates used in industry

Methyl-IC
n-Propyl-IC
i-Propyl-IC
n-Butyl-IC
i-Butyl-IC
t-Butyl-IC
Stearyl-IC*
Cyclohexyl-IC
Phenyl-IC
3-Chlorophenyl-IC
4-Chlorophenyl-IC
3,4-Dichlorophenyl-IC
2,6-Dichlorophenyl-IC
3-Chloro-4-methoxyphenyl-IC
m-Tolyl-IC
p-Tolyl-IC
3-Chloro-4-methylphenyl-IC
4-Methylphenyl-sulfonyl-IC
3-Trifluoromethyl-phenyl-IC
4-Isopropylphenyl-IC
4-Cumolphenyl-IC
1-Naphthyl-IC

* Usually mixture of 90% octadecyl- and 10% hexadecyl-IC

Table 2. Estimated world mono-IC capacity and consumption by region

	Capacity as of 1986 (tonnes)	Consumption as of 1985 (tonnes)
United States	22 000	9 800
Western Europe	46 000	27 400
Japan	2 000	1 400
Total	73 000	38 600

* Includes 3000 tonnes from a mono-IC plant in Israel (1986)

Table 3. Estimated US-consumption of mono-ICs (1985)

Mono-IC	Metric tonnes	Percent
Methyl-IC	7300	75
n-Butyl-IC	1800	18
Aromatic mono-ICs	100	1
Others	600	6
Total	9800	100

Table 4. Estimated western european consumption of mono-ICs

	Metric tonnes	Percent
Methyl-IC	4 000	14
Other aliphatic mono-ICs	500	2
3,4-dichlorophenyl-IC	18 600	68
Other aromatic mono-ICs	3 000	11
m-Trifluoromethylphenyl-IC	1 300	5
Total	27 400	100

Table 5. Companies producing mono-ICs (in alphabetical order)

BASF AG, Federal Republic of Germany
BAYER AG, Federal Republic of Germany
CIBA-GEIGY AG, Switzerland
DOW CHEMICAL, USA
E.I. DUPONT DE NEMOURS & COMPANY INC., USA
HOECHST AG, Federal Republic of Germany
MAKHTESHIM-AGAN, Beer-Sheva, Israel
MITSUBISHI CHEMICAL INDUSTRIES LTD., Japan
RHONE-POULENC SA., France
STAVELEY CHEMICAL INDUSTRIES LTD., England
THE UPJOHN COMPANY, USA
UNION CARBIDE CORPORATION, USA

3.2 Uses

Many of the manufacturers listed in Table 5 have a significant captive consumption of mono-ICs for their pesticides. Products of the *carbamate* type which are obtained by the reaction of aliphatic mono-ICs, especially methyl-IC with OH-groups bearing substances, are well-known insecticides such as,

– Aldicarb: 2-methyl-2-(methylthio)-propylidene-amino-methylcarbamate,
– Carbaryl: 1-naphthalenyl -methylcarbamate,
– Carbofuran: 2,3-dihydro-2, 2-dimethyl-7-benzofuranyl-methyl-carbamate,
– Propoxur: 2-(1-methylethoxy) phenyl -methylcarbamate.

Butyl-IC is the main intermediate for Benomyl, a systemic fungicide and is also used in the production of diabetes medications.

Active substances of the *urea* type are mainly produced on the basis of aromatic mono-ICs, especially 3,4- dichlorophenyl-IC by combining them with primary or secondary amino groups. They play an important role as herbicides, e.g.

– Diuron: *N*-(3,4- dichlorophenyl)-*N*, *N*′-dimethylurea,
– Linuron: *N*-(3,4-dichlorophenyl)-*N*-methoxy-*N*′-methylurea,
– Neburon: *N*-butyl-*N*′-(3,4-dichlorophenyl)-*N*-methylurea.

Fluometron is an important herbicide based on *m*-trifluoromethyl phenyl-IC.

In view of the health hazards which may be involved in using mono-ICs —especially those of high volatility—a number of protective measures have been taken by the industry. Shipping of methyl-IC has been abandoned totally after the Bhopal incident in 1984. In order to eliminate the need for storing larger quantities, pesticide producers generate the methyl-IC in situ and take care of immediate consumption. Closed-loop processes have been developed by which emission of the chemical is avoided. For the transport of mono-ICs in bulk or in drums special containers are provided.

It must be noted that these chemicals are handled almost exclusively by chemical companies whose personnel are well acquainted with the problems of dangerous chemicals. The Bhopal incident has also stimulated efforts to develop alternate routes for the production of carbamates and ureas allowing the abandonment of mono-ICs.

3.3 Characterisation of Selected Monoisocyanates

A thorough literature search was not fully productive to supply physical, toxicological and ecological data for all products listed in Table 1. Its outcome however covers most substances used industrially. Data for some rather insignificant products are included in the following sections.

3.3.1 Physical Data

Table 6 shows physical data for some aliphatic and cycloaliphatic mono-ICs, Table 7 for a number of aromatic mono-ICs.

The figures indicate the considerable differences in the vapor pressure at ambient temperature and the corresponding values of the vapor concentration at the saturation point. These data are important in assessing the vapor inhalation hazard. Densities vary between slightly below and distinctly above the density of water.

The solubility of methyl-IC in water is specified as about 6.7% by weight at 20 °C; about 1.2% of water is soluble in methyl-IC at this temperature [65]. There is no more information available on the solubility of other mono-ICs. Data on the rate of reaction with water are only available for methyl-IC [66], which indicate that the reaction with water vapor is rather slow whereas the reaction in the liquid phase is rapid.

For figures on 2-chloroethyl-IC see [67], for more figures on methyl-IC see (12, 65, 68), on n-propyl-, i-propyl-, n-butyl, i-butyl, t-butyl, stearyl-IC and cyclohexyl-IC see [69], on n-butyl- and cyclohexyl-IC, see also [12].

For more information on phenyl-, 3- and 4- chlorophenyl-, o-tolyl-, 1-naphthyl-IC, see [70], on 3,4-dichlorophenyl and m-tolyl-IC, see [71]. Flash-points [72] and ignition temperatures [73] are presented in Table 8.

Table 6. Physical data pertaining to aliphatic and cycloaliphatic mono-ICs

Mono-IC	Melting point (°C)	Boiling point (°C/mbar)	Density (g/cm^3 20 °C)	Vapor pressure (mbar/°C)	Concentration saturated vapor (g/m^3, 20 °C)
Methyl-IC		38/1013	0.95	513/20	
2-Cl-ethyl-IC		136/1013	1.22	16/25	
n-Propyl-IC	< 30	88/1013	0.90	69/20	235
i-Propyl-IC	< − 75	75/1013	0.87	110/20	374
n-Butyl-IC		115/1013	0.88		
i-Butyl-IC	− 70	102/1013	0.88	33/20	132
t-Butyl-IC	− 75	84/1013	0.85	74/20	300
Stearyl-IC	16	≫ 200*	0.86	10^{-4}/20	0.001
Cyclohexyl-IC	< − 80	171/1013	0.99	7/40	

* Boiling starts at 352 °C

Table 7. Physical data pertaining to aromatic mono-ICs

	Melting point (°C)	Boiling point (°C/mbar)	Density (g/cm^3 20 °C)	Vapor pressure (mbar/°C)	Concentration saturated vapor (g/m^3, 20 °C)
Phenyl-IC	− 31	165/1013	1.095	2.5/20	12
Tri-F-methyl-IC		173/1013	1.34		
3-Cl-phenyl-IC	− 5	201/1013	1.27	0.3/20	2.2
4-Cl-phenyl-IC	27.4	204/1013	1.26	0.3/20	2.1
3,4-DiCl-ph-IC	41	240/1013	1.39	< 0.1/20	< 0.75
o-Tolyl-IC	− 16	189/1031	1.07	0.8/20	4.2
m-Tolyl-IC	− 72	189/1013	1.06	0.7/20	3.6
1-Naphthyl-IC	1	277/1013	1.18	13/127	1.6

Table 8. Flash points and ignition temperatures of various mono-ICs

Mono-IC	Flash point (°C)	Ignition temperature (°C)
Methyl-IC	< − 20	561
n-Propyl-IC	− 1	
i-Propyl-IC	− 10	465
n-Butyl-IC	19	425
i-Butyl-IC	6	430
t-Butyl-IC	− 6	485
tearyl-IC	175	340
Cyclohexyl-IC	53	390
Phenyl-IC	51	645
3-Chlorophenyl-IC	82	695
4-Chlorophenyl-IC	92	650
3,4-Dichlorophenyl-IC	123	650
o-Tolyl-IC	73	615
m-Tolyl-IC	67	615
1-Naphthyl-IC	135	500

3.3.2 Toxicity – Data from Animal Tests – Experience on Humans

Mono-ICs are, as outlined before, handled exclusively within the chemical industry. It is this industry which is faced with the problem of protecting its employees against exposure in the workplace and to take the necessary safeguarding steps. A material which is used solely as a chemical intermediate in technologically advanced chemical plant is obviously a less urgent case for toxicological studies than a product to which large numbers of people are exposed [74]. It is therefore understandable that with regard to mono-ICs only limited information is available from the scientific literature on acute and long-term toxicity findings. Only a few cases of injuries to humans following massive exposure have been reported.

It is on the other hand not surprising that the tragic incidence in Bhopal in December 1984 initiated a remarkable international effort to explore the toxic potential of methyl-IC under various conditions. It is therefore felt that the toxicology of this material should be dealt with in a separate paragraph [75].

3.3.2.1 Methylisocyanate

Methyl-IC due to its extremely high vapor pressure presents the most critical inhalation hazard of all mono-ICs. A study which was published twenty years before Bhopal yielded inhalation toxicity figures as shown in Table 9, see [76].

Symptoms shown by test animals were lung damage with the formation of lung edema. A concentration of 1 . . . 2 ppm caused a significant irritation of mucous membranes in the human respiratory tract even after only short exposure. A maximum concentration of 0.02 ppm for exposure in the workplace is

Table 9. Methyl-IC – acute and subacute inhalation toxicity

LC_{50} 4h rat	12 mg/m^3	(5 ppm)
LC_{50} 2h rat	50 mg/m^3	(21 ppm)
LC_0 2h rat	5.2 mg/m^3	(2 ppm)
LC_0 4h/d 5 d rat	2.7 mg/m^3	(1 ppm)

suggested; a method for the determination of the concentration of methyl-IC in air is presented [76].

The acute oral toxicity is not extraordinarily high (LD_{50}, rat, oral, approximately 100 mg/kg). A dermal LD_{50} of 1800 mg/kg in rabbits was determined [77]. Liquid methyl-IC if administered to the ear of the rabbit causes reddening, edema, necrosis and perforation within 30 minutes; instillation into the eye leads to permanent damage [78]. Massive intoxication of humans by methyl-IC vapor can result in liver and kidney damage [79].

The animal and human response to methyl-IC is the subject of a paper presented at the American Industrial Hygiene Conference [80]. Subsequently detailed circulars were made available by the suppliers which give information on the properties of this chemical, health risks and appropriate handling procedures [65, 81, 82]. A concentration of 0.02 ppm is recommended by official bodies as the maximum admissible in the workplace [83, 84, 85]. Physical and chemical properties, hazards in use and current legislation in France are the subject of a "Fiche Toxicologique" [86]. A detector tube for detecting methyl-IC in air down to the 1 ppm concentration level is considered to be more suitable rather than relying on the irritant characteristic of the chemical as a hazard warning [87]. The explosion limits are 5.3 . . . 26 % methyl-IC by volume in air [88]. Methyl-IC is included as "toxic" in the list of materials classified as transportation health hazards [89].

This was about the state of published knowledge when the incident in the pesticide plant in Bhopal, India occurred on the night of 2nd/3rd December 1984. It would be beyond the scope of this publication to discuss in detail the tragic events of that night and the following days and weeks. It is a matter of fact that an uncontrolled entry of water into a tank containing a considerable amount of methyl-IC, the presence of an iron catalyst and higher than normal amounts of chloroform and finally an elevated temperature (ambient instead of 0°C) initiated an exothermic reaction. Large quantities of the chemical escaped as a vapor cloud through an open safety valve into the densely populated residential area in the vicinity and caused, within a rather short period, a number of fatalities and injuries unparalleled in the history of the chemical industry. The team which was charged by the company involved with investigating the incident and determining its probable cause has submitted a comprehensive report. It was concluded that the incident was the result of a unique combination of unusual events [90]. Authors from India have worked out a model simulation of the methyl-IC dispersion scenario [91].

The symptoms exhibited by the victims and the need for appropriate therapy to treat survivors initiated a unique international effort to study the various toxicological and medical aspects of massive methyl-IC exposure and to clarify whether assumptions (e.g. persistent blindness following cornea corrosion, formation of hydrogencyanide as methyl-IC metabolite) are valid. The main objectives were to study [92]:

– immediate effects of acute exposure
– sensory and pulmonary irritation and injury
– effects on hemoglobin
– effects on other blood constituents and tissue enzyme activities
– pulmonary changes
– effects on the immune function, host resistance and sensitization
– effects on fertility and reproduction
– genetic toxicity

Numerous publications by research teams in Belgium, England, India, USA and Germany have come out in the meantime: more are expected. A conference was organised by the US Department of Health and Human Services on 12 and 13 March 1986. For the papers presented, see [93]. The results communicated in these and other publications have been summarised and interpreted in a detailed review [92], see also [94]. In a further symposium held in France on September 4th 1986 particularly new findings on the inhalation toxicity of methyl-IC were discussed [95]. A paper from a German author deals with observations made at the site of the incident and presumptions resulting therefrom [96], see also [97].

The interested reader may wish to examine these reviews in which references to numerous relevant publications can be found. Attention is also drawn to the results of ongoing animal studies and to epidemiological investigations involving survivors of the disaster. A number of studies are being performed in India.

It is noteworthy that in contrast to early presumptions

– there has been no indication that hydrogencyanide was formed as a metabolite derived from methyl-IC contributing to the death toll,
– although severe irritation of the eye and corrosion of the cornea occurred many times, no cases of permanent blindness were observed,
– although many pregnant women exposed suffered abortion, there was very limited evidence of malformation in newborn children.

3.3.2.2 Other Aliphatic Monoisocyanates

2-Chloroethyl-IC [67]
LD_{50} oral rat: 396 mg/kg
LD_{50} oral mouse: 630 mg/kg
LC_{50} inhal rat 6h: 30 mg/m^3

n-Propyl-IC
LC_{50} ivn mouse: 56 mg/kg [98]
LD_{50} oral rat: 215 mg/kg [69]
LC_{50} inhal rat 4h: 155 ... 201 mg/m^3 [69]
Skin test rabbit 24 h exposure: corrosive [69]

i-Propyl-IC [69]
LD_{50} oral rat: 218 mg/kg
LC_{50} inhal rat 4h: 499 ... 613 mg/m^3
Skin rabbit 24h exposure: moderate irritation
Eye rabbit: strong irritation

n-Butyl-IC
LD_{50} oral rat: 360 mg/kg [69]
LD_{50} oral mouse gavage: 150 mg/kg [99]
LD_{50} oral rat gavage: 600 mg/m^3 [99]
LD_{50} gav guinea pig: 250 mg/m^3 [99]
LC_{50} inhal rat 4h: 80 mg/m^3 [69]
LC_{50} inhal rat 1h: 430 mg/m^3 [69]
LC_{50} inhal mouse: 680 mg/m^3 [99]; inhalation time not specified
Skin rabbit 15 min exposure: severe corrosion [69]

Repeated inhalative exposure of rats and mice to *n*-butyl-IC (and *m*-trifluoromethylphenyl-IC) during several months at 1.3, 6.1, 11.2 mg/m^3: 6.1 and 11.2 mg/m^3 cause functional and histological changes in various organs. Symptoms at all concentrations disappear within 3 months after ceasing exposure [99].

Eighteen persons were exposed for a few minutes to the vapor of butyl-IC after a small amount of the chemical was spilled in a laboratory. Acute respiratory symptoms occurred requiring medical treatment. The electrocardiagrams of 3 patients displayed disorders; attribution to the inhalation was not fully evident [100].

i-Butyl-IC [69]
LC_{50} inhal rat 4h: 235 ... 246 mg/m^3
Skin test rabbit: strongly corrosive after 15 min

t-Butyl-IC [69]
LD_{50} oral rat: 185 mg/kg
LC_{50} inhal rat 4h: 1593 mg/m^3

Hexyl-IC
Guinea pigs were exposed to an aerosol of a conjugate of ovalbumin with hexyl-IC as the hapten group. Pulmonary hypersensitivity assessed by an increased respiratory rate was developed [38]. The immunoglobulin E antibodies formed in the blood of the test animals reacted with the conjugate. No

reaction of the conjugate with antibodies formed after exposure to p-tolyl-IC was observed [101]. Hexyl-IC was tested as a sensory irritant in mice in order to compare the time and concentration dependent response level (decrease of respiratory rate) with phenyl-IC, o-tolyl-IC, p-tolyl-IC and 1,6-hexamethylene diisocyanate (HDI). The aromatic mono-ICs were slightly less potent than HDI; hexyl-IC was the least potent [102].

Stearyl-IC [69]
LD_{50} oral rat: > 5000 mg/kg
Inhal time saturation test, 7h exposure, 14d observation: no symptoms

Octadecyl-IC [103]
LD_{50} ivn mouse: 100 mg/kg

Cyclohexyl-IC
LD_{50} oral rat: 350 mg/kg [69]
LD_{50} dermal rat: 500 mg/kg [69]
LD_{50} ivn mouse: 18 mg/kg [104]
LD_{50} ip mouse: 13 mg/kg [105]
Inhal time saturation test, 10 min exposure: 100% fatalities [69]

3.3.2.3 Aromatic Monoisocyanates

Phenyl-IC
LD_{50} oral rat: appr. 890 mg/kg [70], see also [98]
LD_{50} dermal rabbit, 24h exposure: 7130 mg/kg [77]
LC_{50} inhal rat 1h: appr. 70 mg/m^3 [70]
Skin test rabbit, 24h: corrosive [70, 89]
Eye test rabbit: strong irritation [70, 89]
High mutagenicity resulted when tested on *Salmonella Typhimurium* [106]

Phenyl-IC is found as a trace impurity in commercial polymeric diphenyl-methane diisocyanate (MDI, see 4.1.2.2) in a concentration below 0.05%. There is no evidence that this extremely small amount causes a specific health hazard when processing MDI. As MDI is used in foundries as a core binder, it is not surprising that phenyl-IC is generated by the thermal degradation of the binder [107].

3-Chlorophenyl-IC
LD_{50} oral rat: 3000 mg/kg [70]
LD_{50} dermal rat: > 3000 mg/kg [70]
Skin test rabbit: irritant [70]
Eye test rabbit: irritant [70]

1 mg/m^3 was found to be the threshold concentration for humans; 0.5 mg/m^3 is recommended as the maximum permissible concentration level in the work

place [108]. 0.005 mg/m^3 is suggested as the average permissible concentration in residential areas [109].

4-Chlorophenyl-IC
LD$_{50}$ oral rat: 4710 mg/kg [70, 98]
LD$_{50}$ oral mouse: 530 mg/kg [98, 110]
LD$_{50}$ dermal rat 4h: > 1000 mg/kg [70]
LC$_{50}$ inhal rat 4h: 113 . . . 273 mg/m^3 [70]
Skin test rabbit 24h: irritant [70]
Eye test rabbit: strong irritation to corrosion[70]
Time saturation test (14 days observation after treatment)[70]:

– 3 min exposure: no mortality
– 10 min exposure: 60% mortality
– 30 min exposure: 100% mortality

0.8 mg/m^3 has been found as the irritation threshold limit for humans [108]; 0.5 mg/m^3 is recommended as the maximum level in the workplace [108, 111]; 0.0015 mg/m^3 is regarded as the maximum permissible concentration level in residential areas [109]. Subchronic inhalation studies on rats (4th daily for 4 months at concentrations of 0.1 . . . 0.6 mg/m^3 showed no pathological, behavioral or physiological changes [112].

3,4-Dichlorophenyl-IC
LD$_{50}$ oral rat: 9920 mg/kg [71]
LD$_{50}$ dermal rat 4h: > 1000 mg/kg; no symptoms at 1000 mg/kg [71]
LC$_{50}$ inhal rat 4h: 340–450 mg/m^3 [71]

Minimum lethal conc inhal rat 2h; mouse 4h; 140 mg/m^3; fatalities occurred 10 . . . 12 days (rat) and 1 month (mouse) after exposure [113]
Skin test rabbit 8h: irritant [71]
Eye test rabbit: irritant [71]

m-Trifluoromethyl-phenyl-IC [99]
LD$_{50}$ oral rat gavage: 975 mg/kg
LD$_{50}$ oral mouse gavage: 975 mg/kg
LD$_{50}$ guinea pig gavage: 478 mg/kg
LC$_{50}$ inhal rat: 3600 mg/m^3

For results of subchronic inhalation studies see 3.3.2.2 (*n-butyl-IC*).

o-Tolyl-IC
LD$_{50}$ oral rat: 1700–1900 mg/kg [70]
LD$_{50}$ dermal rat: 1370 mg/kg [70]
Skin test rabbit: corrosive [70]

For results of inhalation studies see 3.3.2.2 (hexyl-IC).
The skin sensitization potential of TDI was weaker than of *m*-tolyl- and *p*-tolyl-IC [114].

m-Tolyl-IC [71]
LD_{50} oral rat: 3000 mg/kg
LC_{50} inhal rat 4h: 30 mg/m^3
For skin sensitization potential see 3.3.2.3 (*o*-tolyl-IC)

p-Tolyl-IC
For the result of inhalation studies, see 3.3.2.2 (hexyl-IC), for the skin sensitization potential, see 3.3.2.3 (*o*-tolyl-IC).

1-Naphthyl-IC
LD_{50} oral rat: > 5000 mg/kg [70]
Time sat test 7h exposure 14d observation: 100% mortality [70]
Skin test rabbit 4h: corrosive [70]
Eye test rabbit: moderate irritation [70]
High mutagenicity occuring in *Salmonella Typhimurium* [106]

3.3.3 Ecological Data

The ecological behavior of mono-ICs is closely linked to their activity in an aquatic environment. As outlined before, miscibility with water is very low. The amount which is dissolvable reacts more or less rapidly; the nature of the reaction products can be manifold. It is therefore extremely difficult to define whether data resulting from tests on fish and microorganisms relate to the isocyanate or its reaction products with water.

Table 10 presents information on the influence of mono-ICs on bacteria [69–71, 115, 116]; Table 11 shows data toxicity toward fish [69–71, 117]. The studies have been carried out in accordance with international guidelines for testing water soluble or emulsifiable substances. These methods have only a very limited utility for water-reactive materials, primarily because a stability of the substance in water of up to 80% for 96 hours is required.

Table 10. Bacteria toxicity of mono-ICs

Mono-IC	EC_0 (mg/l)	Test species	Duration of test (h)	Remarks
n-Propyl-IC	1000	*Pseud Put*	96	Robra
i-Propyl-IC	1000	*Pseud Put*	96	Robra
n-Butyl-IC	1000	*Pseud Put*	96	Robra
t-Butyl-IC	500	*Pseud Put*		Robra
Stearyl-IC	500	*Pseud Fluor*		
Cyclohexyl-IC	36	*Pseud Put*		Bringmann
3-Cl-phenyl-IC	1000	*Pseud Put*		Robra
o-Tolyl-IC	1000	*Pseud Put*		Robra
m-Tolyl-IC	1000	*Pseud Put*		Robra

Pseud Put: Pseudomonas Putida
Pseud Fluor: Pseudomonas Fluorescens

Table 11. Fsh toxicity of mono-ICs

Mono-IC	LC$_0$ (mg/l)	Test species	Duration of test (h)
n-Propyl-IC	100	Brach Re	96
i-Propyl-IC	9.5	Brach Re	96
n-Butyl-IC	100	Leuc Id	96
t-Butyl-IC	49	Brach Re	96
Stearyl-IC	1000	Leuc Id	48
Cyclohexyl-IC	0.5	Leuc Id	72
Phenyl-IC	50	Leuc Id	48
3-Cl-phenyl-IC	100	Brach Re	96
4-Cl-Phenyl-IC	10	Leuc Id	48
3,4-Cl-Phenyl-IC	50	Leuc Id	48
o-Tolyl-IC	100	Brach Re	96
m-Tolyl-IC	19.5	Brach Re	96

Brach Re: Brachydanio Rerio
Leuc Id: Leuciscus Idus

According to [118] the toxicity of 4-chlorophenyl-IC on bacteria is rather high (EC$_{50}$ ≈ 2.6 mg/1).

With regard to the fate of spilt phenyl-IC and 4-chlorophenyl-IC it was found that they are easily oxidizable in water by Cr^{6+}- and Mn^{7+}- Ions; oxygen consumption is very close to the theoretical value [119]. The mutagenic and genotoxic potential of an industrial waste from a pharmaceutical manufacturer containing significant concentrations of these two mono-ICs was tested against *Salmonellatyphimurium* and *Escherichia coli*. Whereas no mutagenicity was detected in the *Salmonella* assay, the wastes proved to be genotoxic in the *E. coli* test [120].

The massive emission of gaseous methyl-IC following the Bhopal incident enabled observation of the behavior of vegetation in the vicinity of the plant toward the chemical. The leaves of a number of plants wilted totally, others only partially. A limited number of plants were found to be resistant. Healthy leaves developed in the following spring but the flowering and fruiting appeared not to be normal in most of the species [121]. The chlorophyll damage which took place in various plants was more than 25% on 26 plant species examined at a distance of 500 m from the emission source [122]; see also [123]. Limnochemical and water quality studies at a lake in the Bhopal area in the winter season immediately after the incident showed pollution by nitrates, chlorides and thiocyanates. The results can only be interpreted with difficulty [124].

4 Polyfunctional Isocyanates [125] –
Technical Data – Application

As outlined in Section 1 diisocyanates and isocyanates containing higher functionality (most of them derived from diisocyanates) represent more than 95% of
the total market for this group of chemicals. In Section 2 their key function as
the basis of an important class of polymers – polyurethanes – has been illustrated. These products are made by companies in various sectors outside the
chemical industry. Procedures for the non-industrial generation of polyurethanes are also very well developed; do-it-yourself uses are common practice.
This section surveys production capacities, physical and chemical properties
and areas of application for commercial polyisocyanates.

It is obvious that numerous individuals all around the world handle these
chemicals at work or outside work. It is the distinct responsibility of the
isocyanate suppliers to make the full information available to the user on the
hazards which may occur due to improper handling of isocyanates and to give
full guidance on the necessary precautionary measures. These aspects as well as
experience on the interaction of these substances with the environment are
specifically dealt with in Sections 5, 6 and 7.

4.1 Diisocyanates

Toluene diisocyanate (TDI) and diphenylmethane diisocyanate (MDI) – including the "polymeric" grade of this product, see 4.1.2.2 – have by far the greatest
importance in the market place. Hexamethylene diisocyanate (HDI), isophorone
diisocyanate (IPDI), dicyclohexylmethane diisocyanate (HMDI) and naphthy-
lene diisocyanate (NDI) as well as trimethyl hexamethylene diisocyanate
(TMDI), phenylene diisocyanate (PPDI), cyclohexyl diisocyanate (CHDI), xylylene diisocyanate (XDI) and 1,3-bis(isocyanatomethyl) cyclohexane (HXDI) find
a variety of special applications.

Most of the aforementioned abbreviations which are used in the following
sections are customary in the industry.

4.1.1 Capacities and Consumption

A market survey published some years ago [5] presents detailed figures for TDI
and MDI, see Table 12. For the aliphatic and cycloaliphatic types only summary
data are available, see Table 13. No information is at hand about the remaining
aromatic diisocyanates.

Table 12. Toluene diisocyanate (TDI) and diphenylmethane diisocyanate (MDI) capacity, consumption, export (1985)

Figures in thousand metric tonnes

	Year end production capacity	Domestic consumption	Export
TDI			
United States	288	224	59
Western Europe	369	182	110
Japan*	78	72	15
Total	735	478	
MDI			
United States	365	246	54
Western Europe	414	305	53
Japan	118	79	17
Total	897	630	

* Plant capacities increased during 1985, although increases were not announced officially

Table 13. Aliphatic and cycloaliphatic diisocyanates – 1985 consumption

Figures in metric tonnes

United States	13 600
Western Europe	20 000
Japan	3 000
Total	36 600

The authors of [5] project the following average annual growth rates for 1985 through 1990:

– TDI: USA 1.3%, Western Europe 1.5 . . . 2%, Japan 2%;
– MDI: USA 3%, Western Europe 3 . . . 4%, Japan 6 . . . 8%;
– aliphatic and cycloaliphatic diisocyanates: 6.5 . . . 9.4%.

It should be noted that the aforementioned Table and figures only include capacities, consumption and export data for the USA, Western Europe and Japan. According to Table 14, in which the producers of diisocyanates are listed, TDI and MDI are also produced in Latin America, Korea and in several socialist countries.

Figure 1 shows that in 1987 the world capacity for production of TDI and MDI was close to 2 million metric tonnes [126]. It is noteworthy that polymeric grade MDI represents more than 90% of the total MDI market.

Table 14. Companies producing diisocyanates classified by region (alphabetical order)

United States
BASF Corporation (TDI, MDI)
DOW CHEMICAL USA (TDI, MDI)
ICI AMERICA INC. (TDI, MDI)
MOBAY CHEMICAL CORPORATION (TDI, MDI, others)
OLIN CORPORATION (TDI)

Latin America
BAYER, Brazil (MDI)
CYDSA, Mexico (TDI)
PETROCHIMICA RIO TERCERO, Argentina (TDI)
PRONOR, Brazil (TDI)

Western Europe
AKZO CHEMIE BV, Holland (miscellaneous)
BASF AG, Federal Republic of Germany (MDI); production in Belgium
BAYER AG, Federal Republic of Germany (TDI, MDI, others); production in FRG, Belgium, Spain
DOW CHEMICAL EUROPE, Switzerland (MDI); production in Portugal
HÜLS AG, Federal Republic of Germany (miscellaneous)
ICI EUROPA LTD., Belgium (MDI); production in Holland and England
MONTEDIPE SPA, Italy (TDI, MDI)
RHONE POULENC SA, France (TDI, miscellaneous)
SAPICI SPA, Italy (miscellaneous)

Eastern Europe
SACHEM, Poland (TDI)
SODA So, Yougoslavia (TDI)
State Enterprise, USSR (TDI, MDI)
VEB SYNTHESEWERK SCHWARZHEIDE, German Democratic Republic (TDI, MDI)

Far East
CHINYANG CHEMICAL, South Korea (TDI)
MITSUI TOATSU CHEMICAL INC., Japan (TDI, MDI)
NIPPON POLYURETHANE INDUSTRY CO LTD., Japan (TDI, MDI)
STATE ENTERPRISE, Peoples Republic of China (TDI, MDI)
SUMITOMO BAYER URETHANE CO LTD., Japan (TDI, MDI, others)
TAKEDA CHEMICAL INDUSTRIES LTD., Japan (TDI)

4.1.2 Physical and Chemical Data

The physical state of diisocyanates at ambient temperature and their viscosity
are important factors in facilitating easy processing; the vapor pressure essentially determines the exposure hazard. The bulk of commerical products (TDI,
polymeric MDI) are liquid and of low viscosity. The vapor pressures of
diisocyanates are rather different but their volatility is generally low in contrast
to most monoisocyanates.

4.1.2.1 Toluene Diisocyanate – TDI

Commercial TDI is usually a mixture of the 2,4- and 2,6-isomer. Its synthesis
starts from toluene which is transformed into a mixture of 2,4- and 2,6-
dinitrotoluene. After reduction and subsequent phosgenation, a mixture of 2,4-

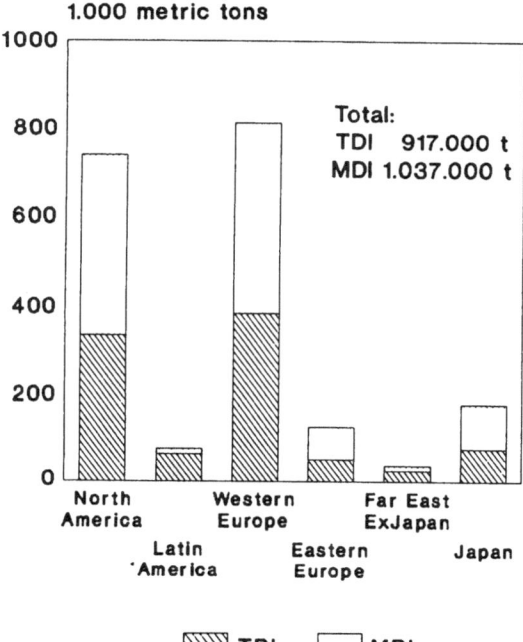

Figure 1. TDI and MDI Capacities 1987

and 2,6-toluene diisocyanate is obtained which is purified by distillation. The isomer ratio can be varied within certain limits.

The main sales product is TDI 80 which contains 80% of the 2,4-isomer. TDI 65 (65% 2,4) and TDI 100 (the pure 2,4-isomer) are supplied to a limited extent. The pure 2,6-isomer is not commercially available.

TDI of a standard quality is marketed by the various suppliers. In order to ensure reproducible processing characteristics, it is important to keep the total and hydrolysable chlorine content – which originates from the production pro-cess – low and constant.

Properties of TDI 100, TDI 80 and TDI 65 are shown in Table 15. For more infomation see (127–132); methods for analytical quality control are described in [24], chromatographic methods to determine TDI in prepolymers are reported in [133, 134, 135].

4.1.2.2 Diphenylmethane Diisocyanate – MDI

The basis of industrial MDI production is the phosgenation of the mixture of amines obtained by the condensation of aniline and formaldehyde. It is commer-cially available in two forms, namely the "monomeric" and the "polymeric" type. Monomeric MDI is mainly 4,4'-diphenylmethane diisocyanate which usually contains small amounts of the 2,4'- and traces of the 2,2'-isomer.

Table 15. Toluene diisocyanate (TDI); physical and chemical properties*

Molecular weight	174.2
NCO-content	48.3%
Appearence	Mobile colourless to pale yellow liquid
Odor	Sharp, pungent
Specific gravity	1.22 g/ml at 25 °C
Melting point	TDI 65 < 8 °C
	TDI 80 < 15 °C
	TDI 100 22 °C
Boiling point	251 °C at 1013 mbar
Vapor pressure	0.021 mbar at 20 °C
Viscosity	0.03 mPa.s
Flash point	135 °C (DIN 51376)
Autoignition temperature	666 °C (DIN 51794)
Explosion limits	Upper Lower
	0.9%v/v at 118 °C 9.5% v/v at 150 °C
Conversion factors	1 ppm $\approx$ 1ml/m^3 $\approx$ 7.23 mg/m^3

* Data are valid for TDI 65, TDI 80 and TDI 100 if not indicated otherwise

Polymeric MDI is, in contrast to the monomeric type, a nondistilled product. Its main constituent is diphenylmethane diisocyanate; the remainder is a combination of isocyanates of higher functionality. There are several types of polymeric MDI on the market with varying viscosities and different contents of the 2,4-isomer and isocyanates of higher functionality. They usually contain phenylisocyanate as a trace impurity (see 3.3.2.3.1); a method for chromatographic determination is described in [136].

Although the functionality of polymeric MDI is > 2 it seems to be reasonable to discuss it together with the diisocyanates.

For the properties of MDI see Table 16 and [128, 137, 138]. The evidence brought forward to show that MDI forms a hydrate in an aqueous environment seems to be questionable [139].

The analytical control of the MDI content in prepolymers can be carried out according to [134, 135].

The extremely low vapor pressure at ambient temperature minimizes the exposure risk and is therefore an important feature in its application [140].

4.1.2.3 Other Diisocyanates

1,5-Naphthylene Diisocyanate (NDI)

NDI is a special product for the manufacture of polyurethane elastomers. Due to its high melting point, it has to be processed at rather elevated temperatures requiring particular care. The same is essential for handling the solid chemical (powder or flake). For its properties see Table 17 and [141].

Table 16. Diphenylmethane diisocyanate (MDI); physical and chemical properties

	Monomeric MDI	Polymeric MDI*
Molecular weight	250.3	320...400
NCO-Content	33.6%	$\approx$ 30...31%
Appearence	Fused solid or flakes	Viscous liquid
	White-pale yellow	Dark amber
Odor	Slightly musty	Slightly musty
Specific gravity	1.23 g/ml at 25 °C	1.23 g/ml at 25 °C
Melting point	38 °C	Not applicable
Vapor pressure at 25 °C	< 0.00001 mbar	< 0.00001 mbar
Viscosity	5 mPa.s at 50 °C	100 ... 250 mPa.s at 25 °C
Flash point (ASTM D 93)	202 °C	> 200 °C
Autoignition temperature	240 °C (DIN 51794)	> 200 °C (DIN 51794)
Conversion factor	1 ppm $\approx$ 1 ml/m^3 $\approx$ 10.4 mg/m^3	

* Selected values: properties vary with commercial products.

Table 17. 1,5-Naphthylene diisocyanate (NDI); *p*-Phenylene diisocyanate (PPDI) *trans*-1,4-Cyclohexyl diisocyanate (CHDI); physical and chemical properties

	NDI	PPDI	CHDI
Molecular weight	210.2	160.1	166.2
NCO-Content	40%	51.4% min.	49.5% min.
Description	White flake	White flake	White flake
Odor	Aromatic	Pungent	Pungent
Specific gravity	1.42 g/ml/20 °C*	1.17 g/ml/100 °C	1.12 g/ml/70 °C
Melting point	127 °C	94 ... 95 °C	58 ... 62 °C
Boiling point	Not applicable	260 °C/1013 mbar	260 °C/1013 mbar
Vapor pressure	< 0.0001 mbar/20 °C	0.011 mbar/20 °C	0.005 mbar/20 °C
Viscosity	Not applicable	1.17 Pa.s/100 °C	13 mPa.sec/65 °C
Flash point	192 °C (DIN 51758)	> 93 °C (PMCC)**	> 93 °C (PMCC)**
Conversion factor	1 ppm $\approx$ 8.7 mg/m^3	1 ppm $\approx$ 6.6 mg/m^3	1 ppm $\approx$ 7.0 mg/m^3

* Molten and resolidified product
** Pensky-Martin-Closed-Cup

1,4-Phenylene Diisocyanate (PPDI)

PPDI which bears the two isocyanate groups in 1,4-position has a limited importance for the manufacture of elastomers. Its physical status also requires very careful handling. For its properties see Table 17 and [142].

Cyclohexane 1,4-Diisocyanate (CHDI)

CHDI (*trans* configuration) is a special isocyanate used for producing elastomers. For its properties see Table 17 and [143].

1,6-Hexamethylene Diisocyanate (HDI)

HDI is a key product in the synthesis of polyisocyanates having an extremely low (< 0.5%) content of free (monomeric) HDI which are used as components

Table 18. Hexamethylene diisocyanate (HDI); trimethylhexamethylene diisocyanate (TMDI); xylylene diisocyanate (XDI); physical and chemical properties

	HDI	TMDI	XDI
Molecular weight	168.2	210.3	188.2
NCO-Content	50%	39.8%	44.6%
Appearence	Colorless to sl. amber liquid	Colorless to lt. yellow liquid	Colorless liquid
Specific gravity	1.05 g/ml/20 °C	1.01 g/ml/20 °C	1.2 g/ml/20 °C
Melting point	− 67 °C	− 80 °C	7.5–12.5 °C
Boiling point	255 °C/1013 mbar	149 °C/13 mbar	151 °C/8 mbar
Vapor pressure	0.011 mbar/20 °	0.001 mbar/20 °	0.008 mbar/20 °
Viscosity	2.5 mPa.s/20 °C	5 mPa.s/20 °C	3.6 cps/20 °
Flash point	135 °C (DIN 51758)	148 °C (DIN 51758)	185 °C (PMOC)*
Autoignition Temp.	460 °C (DIN 51479)		
Conversion factor	1 ppm ≈ 7.0 mg/m^3	1 ppm ≈ 8.3 mg/m^3	1 ppm ≈ 7.8 mg/m^3

* Pensky Martin open cup

for light-stable, reactive polyurethane coatings. No specific hazard from exposure to the monomer does therefore exist for the user. Table 18 shows the properties, see also [144].

Trimethyl Hexamethylene 1,6-Diisocyanate (TMDI)
Commercial TMDI is a mixture of 2,2,4- and 2,4,4-trimethyl hexamethylene diisocyanate; the isomer ratio is approximately 50: 50. The product is mainly used for special light-stable and weather-proof coatings. For the properties see Table 18 and [145].

Xylylene Diisocyanate (XDI)
Commercial XDI is a mixture of ca. 70% *m*- and 30% *p*-xylylene diisocyanate. It is a speciality chemical for light-stable coatings. For its properties see [146] and Table 18.

Dicyclohexylmethane 4,4'-Diisocyanate (HMDI)
HMDI (usually supplied as a mixture of several steric configurations) is a special product with growing interest for the coatings industry. For its properties see Table 19, see also [147].

Isophorone Diisocyanate (IPDI)
IPDI (3-isocyanatomethyl 3,5,5-trimethylcyclohexyl isocyanate) is a special product for the synthesis of polyisocyanates for the coatings and related industries and has also found a limited use for the production of polyurethane plastics. For its properties, see [148, 149] and Table 19.

1,3-Bis (isocyanatomethyl) Cyclohexane (HXDI)
HXDI is sold mainly for the synthesis of non-yellowing 2-component coatings. Its properties are described in [150, 151], see also Table 19.

Table 19. 4,4'-Dicyclohexylmethane Diisocyanate (HMDI); *iso*phorone diisocyanate (IPDI); 1,3-bis (*iso*cyanatomethyl) cyclohexane (HXDI); physical and chemical properties

	HMDI	IPDI	HXDI
Molecular Weight	262.3	222.3	194.2
NCO-Content	32%	37.8%	43.2
Description	Colorless liquid	Slightly amber liquid	Colorless liquid
Odor	Slight, lachrymating	Pungent	Slight
Specific Gravity	1.07 g/ml/25 °C	1.06 g/ml/20 °C	1.01 g/ml/25 °C
Melting Point	19–23 °C	$-60\,°C < -50\,°C$	
Boiling Point		158 °C/13 mbar	154 °C/13 mbar
Vapour Pressure	0.001 mbar/25 °C	0.0004 mbar/20 °C	0.011 mbar/20 °C
Viscosity	30 mPa.s/25 °C	5.8 cps/25 °C	
Flash Point	200 °C (DIN 51758)	155 °C (DIN 51758)	150 °C (PMOC)*
Autoignition Temp.	430 °C (DIN 51479)		
Conversion Factor	1 ppm ≈ 10.9 mg/m^3	1 ppm ≈ 9.2 mg/m^3	1 ppm ≈ 8.1 mg/m^3

* Penski Martin Open Cup

4.1.3 Application

Diisocyanates are mainly used as reaction partners for compounds bearing active hydrogen atoms (predominantly polyols); the reaction conditions can be varied within very wide limits by using catalysts and, if appropriate, by the application of heat. The time which elapses between combining the reactants and attaining the state of curing can vary between seconds and hours. If an equimolecular amount of isocyanate is used, the resulting polymer usually has a high molecular weight; after completion of the reaction (approx. 24 hours) the final product does not contain monomeric diisocyanate any more.

The use of excess isocyanate leads on the other hand to lower molecular weight adducts ("prepolymers" or "semi-prepolymers") which contain free NCO-groups and possibly – depending on the relation of the reactants – monomeric diisocyanate as well. The latter predominates if diisocyanates are "modified" by the incorporation of small amounts of reactive components. The designation "polyisocyanate" is customary for prepolymers having more than two NCO-groups within the molecule [125]. Polyisocyanates are also obtained by dimerization or polymerization of the monomer. The combination of these products with suitable active components leads in a second step to the final polymer (see also Section 4.2.1).

The compliance with the given stoichiometric relation of the ingredients and efficient mixing are important prerequisites for obtaining a final product of high quality. For the industrial processing of the liquid reactive systems in the manufacture of polyurethane [152] polymers, sophisticated technologies have been developed allowing continuous or discontinuous metering and mixing of the components in a very wide output range. Devices for the formation of polyurethanes outside industrial sites and for non industrial applications are more simple and mostly mobile. In the do-it-yourself area moisture curing so called one-component-polyurethane foam can be dispensed from small cans.

Polyisocyanates, prepolymers, semi-prepolymers and modified isocyanates are mostly produced by the chemical industry and supplied to the end-user. The following paragraphs outline the various application areas for diisocyanates. A detailed and comprehensive survey is given in [153]; this review includes information on processing equipment, properties of end products and occupational hygiene as well as environmental aspects. A more recent publication covers the entire field [154] of processing and application. Attention is also drawn to the proceedings of various international conferences dealing with this subject [155, 156].

4.1.3.1 Cellular Polyurethanes

TDI and MDI are the predominant diisocyanates for the manufacture of cellular polyurethanes. Polyethers and to a small extent polyesters are the main reaction partners; glycols or aromatic amines may be used as chain extenders; tertiary aliphatic amines and organic tin compounds are used as catalysts. Polyether siloxanes act as surfactants and foam stabilizers. The expansion of the raw material mixture is effected by carbon dioxide gas generated by the reaction of TDI with water added to the system in well defined quantities. Highly volatile organic solvents (especially chlorofluorocarbons) are used as additional blowing agents for foams with open cells. They are on the other hand also the main blowing agent for closed cell materials.

Flexible polyurethane foam is manufactured as slabstock and as moulded article. TDI is the principal diisocyanate for slabstock which is produced in numerous plants all over the world on continuously operating machines with an output of up to 500 kg/min. Slabstock dimensions of 2 m wide and up to 1.20 m high are the state of engineering. Curing of the slab takes approximately 24 hours at room temperature [157]. Commercial foams have densities between 15 to 40 kg/m^3; their hardness and cell structure can be varied within certain limits.

The slab is further processed by cutting it into sheets, foils and shaped articles. The furniture and matress industry are the main consumers of material from slabstock; application in the automotive and textile industries is considerable.

Moulded flexible foam articles have achieved increasing importance as seating material for vehicles and to some extent for furnishings. Discontinuously operating machines allow the precise control of the amount of reaction mixture which is needed to fill the volume of the respective mould after expansion. In order to avoid adhesion of the rising foam, the mould has to be treated with efficient release agents. Depending on the type of polyol, heating of the mould to 100 . . . 120 °C after introduction of the components is necessary to obtain a proper production cycle and acceptable foam properties. TDI is used for this "hot cure" process. For the "cold cure" technique, mainly mixtures of TDI and MDI are employed in combination with specific polyethers. The resulting foam is "highly resilient" ("HR-foam"); mouldings of this type are widely used in the automotive industry.

Rigid polyurethane foam is characterized by a closed cell structure. The water absorption is therefore very low. Due to the presence of chlorofluoro-carbon vapor in the cells, the thermal insulation efficiency reaches a level which is unmatched by other standard insulation materials. Foam densities vary between 30 and approximately 300 kg/m^3. The mechanical strength of the foam and its adhesion to facings are specific features for application in the construction and in the appliance industries.

Polymeric MDI is mainly used for rigid foam in combination with polyols whose molecular structure leads to a high degree of cross-linking thus assuring the desired rigidity.

Applications of rigid foam are manifold:

- domestic appliance industry: insulation of refrigerators and freezers in which the foam is integrated into the whole construction;
- construction industry: insulation slabs particularly for roofs, sandwich elements for walls and roofs, spray foam for the direct insulation of surfaces, one component foam from pressurized cans for fixing building elements;
- technical insulation: prefabricated pipes for district heating, protection of storage tanks and pipes against loss or inflow of heat;
- transportation industry: containers for perishable goods, refrigerated vehicles, interior liners for automobiles, cavity filling in car bodies;
- mining industry: consolidation and strengthening of coal seams and surrounding strata;
- packaging: direct encapsulation of goods, thin sandwiches for cardboard boxes.

It becomes obvious from these examples that the applications of rigid foam are widespread, the reason why handling of the chemicals should be as simple as possible. For many uses 2-component systems are available which makes processing very easy.

4.1.3.2 Microcellular Polyurethane Elastomer Mouldings

The designation "Microcellular Elastomer" covers a range of products produced by the so called "Reaction Injection Moulding" (RIM) process; the terms "Integral Skin" or "Structural Foam" are also customary for these moulded parts which are characterized by a dense non-cellular skin and a microcellular core. Their properties are not only determined by the polyurethane matrix but also by the density distribution over the cross section. Reaction systems based on modified MDI in combination with special polyether polyols (polyesters in special cases) are used for the production of these articles which, depending on the polyol, can be flexible or rigid. IPDI is employed to a small extent for light-stable end products.

In order to minimize the cycle time, the reaction rate has usually to be very fast; preheating of moulds which must have a perfect surface is customary. "External" as well as "internal" release agents ensure a good release effect. The components are usually coated before use.

In recent years fiber reinforcement of RIM-mouldings has been developed leading to "RRIM"-products.

The application of microcellular polyurethanes has shown a remarkable growth in the last two decades. Typical uses are:

– automotive industry: head rests, covers for steering wheels, motorbike saddles, car bumpers and spoilers, flexible front ends, fascias, window frames;
– shoe industry: soles for nearly every type of shoe, directly moulded to the uppers or produced individually.
– construction industry: window and skylight frames, roof drains and vents.

4.1.3.3 Solid Polyurethane Elastomers – Elastic Fibers

Solid (or "massive") elastomer articles can be obtained by a casting process (CPU) or on the basis of thermoplastic polyurethanes (TPU) which are processed by conventional techniques.

Casting a reactive mixture into an open mould was one of the first methods developed at the end of the thirties following the invention of the diisocyanate polyaddition process. Practically all commercial diisocyanates in combination with polyesters and polyethers as well as under different curing conditions (ambient or elevated temperature – "cold" or "hot" curing) have been studied since that time.

The oldest high property, hot cure casting system is based on NDI and a linear polyester. To date, most CPU-systems are based on TDI (as prepolymer) and monomeric MDI. HDI, IPDI, PPDI and CHDI are used only for very special formulations. Diols or primary aromatic diamines assure a true chemically cross-linked network.

Polyurethane elastomers with thermoplastic character (TPU) are based mainly on MDI which is combined with a polyether or polyester and a low molecular weight glycol for chain extension. HDI and IPDI are used in specific cases. The mixture of the chemicals is dispensed on a moving belt where upon it solidifies. The sheets are then ground in a granulator.

Elasticity over a wide temperature range is a common characteristic of all TPU- and CPU-elastomers regardless of whether they are soft or very hard. They are particularly distinguished by their excellent abrasion-, wear- and weather-resistance, stability to light, atmospheric ozone and UV-radiation as well as the absence of extractible ingredients (like plasticizers) and good resistance to oil, grease and many solvents. Numerous applications which make use of these properties have been developed. Examples are:

– transportation: joints, grease boots, membranes, bearings, shock absorbing elements, car body parts;

– general Engineering: rollers, roll covers, milling rolls, damping elements, drive gears, wear resistant sieves;
– construction: sport and track surfaces, pipe seals, formwork mats;
– electrical: cable fittings, encapsulation of switches, cables and capacitors, wire and cable coatings, cable take-outs;
– footwear: ski boots, sport shoe soles.

Elastomeric polyurethane fibers are based mainly on MDI- and to some extent on TDI-prepolymers which are combined with special polyethers or polyesters in conjunction with chain extenders. The reaction is performed in a suitable solvent; thereafter, the solution is extruded through a multi-opening spinneret into a heated chamber. The filament is formed by evaporation of the solvent. In contrast to this "dry spinning procedure" a "wet spinning" and a "reaction spinning" process are also practised.

Such elastomeric fibers are mainly incorporated into fabrics to achieve an improvement in properties such as their elastic recovery after high extension.

4.1.3.4 Modified Diisocyanates

TDI, polymeric MDI, HDI, IPDI, to some extent HMDI, TMDI, XDI and HXDI are mostly used in a modified form as polyisocyanates for polyurethane coatings, adhesives and for binding organic or inorganic particles, although sometimes the original form is used.

4.2 Polyisocyanates for Polyurethane Coatings, Adhesives and Binders

Polyurethane systems based on polyisocyanates have a special place among the many resins used as the polymer component in coatings, adhesives and binders.

Polyurethanes and polyfunctional isocyanates exhibit in situ a pronounced and durable adhesion to the surfaces of many materials. This is mainly responsible for the broad application of these materials as adhesives and binders. Excellent mechanical properties, chemical resistance, weather stability and corrosion protection are important additional features for the success of polyurethane coatings.

4.2.1 Polyurethane Coatings

The following classification of the major product groups is appropriate for polyurethane coatings:

1. Two-component systems in which the polymer is formed by a reaction of a polyisocyanate and a polyol;
2. One-component systems in which the polymer is formed by reaction of a polyisocyanate with atmospheric moisture;

3. One-component systems containing a blocked polyisocyanate and a polyol;
   the blocking agent is split off by applying heat thus liberating the polyisocyanate which reacts with the polyol.

Diisocyanate monomers as mentioned in 4.1.3.4 can be modified by

– conversion into an *adduct* (prepolymer) with a triol such as trimethylolpropane or higher molecular weight polyols;
– allowing them to react with water under specific conditions leading to a *biuret
  derivate*;
– by trimerization or polymerization forming *polyisocyanurate ring structures.*

TDI or mixtures of TDI and HDI are the preferred products for polyisocyanates based on adducts. They are usually marketed as solutions in
ethylacetate or other inert solvents. Their solvent content is in the order of
25 . . . 40% [158,159].

Polyisocyanates of the biuret- and the polyisocyanurate type are normally
derivates of HDI or IPDI. The HDI-based biuret derivative is available solvent-
free or in combination with 10 . . . 30% solvent [160,161], whereas the IPDI-
biuret and the polyisocyanurates are offered as solutions only [162, 163]. In
order to avoid exposure to the vapor of the relatively more volatile diisocyanates (TDI, HDI, IPDI) during the application of the coating, the content of
residual monomers in the product sold must be kept at the technically feasible
minimum. It is always below one percent; most products sold contain less than
0.5 percent monomer. For the gas chromatographic determination of residual
diisocyanate monomer, see [164,165].

The aforementioned two- and one-component coating systems which normally contain polyesters as reaction partners for the isocyanate are solvent-borne.
Curing usually takes place at ambient temperature with the exception of
one-component coatings containing a blocked polyisocyanate. The most common product in this group – the adduct of TDI and trimethylolpropane with
phenol blocking the NCO-groups – is thermally degraded at 160 . . . 180 °C
thus liberating the isocyanate which subsequently reacts with the polyol component [166].

Solvent-borne non-reactive high molecular weight polyurethanes which
form the coating film after physical drying and dry-in-air systems which are
produced by the reaction of diisocyanates with polyol-modified drying oils are
further examples of polyurethane surface finishes applied with the use of organic
solvents.

For solvent-free, two-component and one-component coatings, MDI adducts (prepolymers) based on polymeric MDI [167] in combination with polyethers are the preferred system. TDI- and IPDI-adducts blocked with
caprolactam [168] in combination with special polyesters can be converted to
a dry powder which is utilized for powder coatings.

Polyisocyanate based coatings have found a broad range of application, e.g.
for wood and metal finishes, for corrosion protection, for the surface treatment

of rubber and plastic articles, for the insulation of wires, for the generation of durable self-leveling layers on floors. Their ability to cure at ambient temperature offers significant advantages over hot cure coatings. Systems based on aliphatic diisocyanates show permanent light stability even if they do not contain pigments.

For more information on the chemistry, properties and applications, see [169,170].

A special area of use for reactive polyurethane systems or non-reactive polyurethanes is the refinement and finishing of textiles, paper and leather. These materials have several advantages over synthetic rubber, acrylics, nitrocellulose and PVC. The treatment can be performed in solution, dispersion, or in a low-solvent or solvent-free "high solid" system as well as in solid granulate or powder form. Abrasion and tear resistance as well as tensile strength are high; the hardness can be varied without plasticizer addition by altering the polymer structure. The light stability is excellent if aliphatic isocyanates are used in the systems. Additional advantages are the very good adhesion to many substrates, the high flexibility (even at low temperatures) and the good resistance to solvent.

4.2.2 Polyurethane Adhesives

The various polyurethane adhesive systems can be distinguished as follows:

1. Two-component reaction system adhesives consisting of a low molecular weight polyisocyanate and a rather low molecular weight polyol;
2. One-component reaction adhesives consisting of a higher molecular weight polyisocyanate which reacts with moisture to form the adhesive bond;
3. One-component solvent-borne polyurethane adhesives containing a high molecular weight polyurethane;
4. Two-component solvent-borne adhesives deriving from the one-component polyurethane adhesive to which a polyisocyanate is added as crosslinker;
5. Dispersion adhesives containing a high molecular weight polyurethane dispersed in water.

TDI, MDI and HDI are used for the production of adhesive raw materials. Triphenylmethane 4,4',4''-triisocyanate [171] and thiophosphoric acid-tris-(p-isocyanato phenylester) [172] are used as crosslinker in solvent born two-component adhesives. They are usually supplied as 20 percent solution in dichloromethane.

The TDI adduct yielding trimethylolpropane [158] is used for solvent containing polymeric MDI and for solvent-free, two-component reaction adhesives in combination with polyesters or polyethers. The basis for one-component reaction adhesives are high molecular weight, NCO-terminated, polyurethane prepolymers which are applied in solution and cured by reaction with moisture in the air. Cross linking can be achieved by adding triisocyanates, as mentioned in the preceeding paragraph. Aqueous polyurethane dispersions have advantages over many other dispersion adhesives because they are free of organic solvents and therefore less hazardous.

It should be noted that polyisocyanates, specifically the aforementioned triisocyanates, serve as crosslinkers for adhesives based on other polymers such as polychloroprene, nitrile rubber or natural rubber. Polyurethane adhesives are used in the footwear industry, for bonding plastics, in plastic film composites especially for food packaging, in the clothing industry for the lamination of flexible polyurethane foam foils on textile, in the automotive industry for sealants subjected to high stress, in the building industry for the manufacture of sandwich elements with plastic foam. For further information on their chemistry, properties and applications, see [173].

4.2.3 Binders for Organic and Inorganic Particles

A significant amount of polymeric MDI and MDI-based polyurethane systems is used for binding organic and inorganic particles.

MDI is used in particular for binding forest and agricultural products. Wood particles are the main substrate to date; the production of MDI bonded particle board has achieved significant importance. Processing is carried out on the same equipment and under similar conditions as the production of particle board using urea- or phenol-formaldehyde resins. Curing of the isocyanate bound board requires a temperature in the area of 200 °C, similar to conventional resins. Reduced binder content, increase in mechanical strength and improvement of humidity tolerance are the main advantages of this material [174]. Polymeric MDI is also used as binder for milled rubber waste and foam scrap.

Polyurethane systems based on MDI in combination with modified alkyd or phenolic resins are customary in the foundry industry as binders for sand in the preparation of mold cores. Curing occurs in a few seconds by blowing a stream of amine-catalyst through the core [175].

5 Polyfunctional Isocyanates – Toxicity and Ecotoxicity

The toxicity and ecotoxicity data which are presented in this section are specifically based on the outcome of laboratory studies on test animals, in vitro studies on human and animal body fluids and microorganisms as well as of analytical determinations on substrates whose interaction with the test substances was evaluated. It goes without saying that the exclusive aim of these investigations is to assess the hazard which may occur if humans and the living environment are exposed to these chemicals and their reaction products.

Polyfunctional isocyanates are handled – as outlined in the preceding section – by numerous employees under various circumstances at their places of work. An in depth judgment of their acute, subchronic and chronic toxicity is therefore essential in order to take prophylactic measures for minimizing the exposure. Health damage which can occur accidentially may initiate specific toxicity studies whose outcome could assist in estimating the real health risk.

Accidental spillages during transport and handling of polyfunctional isocyanates cannot be completely avoided; it is therefore essential to have information available on their interaction with air, soil, water and their influence on plants. This data assists one in taking the best possible countermeasures to deal with accidents.

5.1 Toxicity-Data from Animal Tests

An outline of animal test data determined for polyfunctional isocyanates will be given in this section; discussion of occupational safety and health aspects is the subject of Sections 6 and 7, see [75].

Several reviews which have been published since the early fifties describe the respective state of knowledge; they deal mainly with TDI, MDI and HDI as these chemicals are the most important and subsequently most thoroughly investigated products [127–132, 137, 176–184].

5.1.1 Acute Oral Toxicity

The acute oral toxicity of diisocyanates is generally low; LD_{50} figures are mostly in the order of 5000 mg/kg and higher. Only PPDI, HDI and CHDI show somewhat lower figures. A selection of published data is presented in Table 20 to which the following literature references apply:

TDI – a:[176, 185], b: [128]
MDI – c: [186, 187], d: [128]
NDI – [141]
PPDI – [142]
CHDI – [143]
HDI – e: [144], f: [188], g: [189]
TMDI – [145]
XDI – h: [190], i: [146]
HMDI – j: [147], k: [191]
IPDI – l: [192], m: [193]
HXDI – h: [190], i: [146]

Table 20. Diisocyanates – acute oral toxicity

Product	LD_{50} (mg/kg b.w.)	Test species	Remarks*
TDI	5800	Rat	a
TDI	4130	Mouse	b
MDI	> 10000	Not specified	c mono- and polymeric MDI
MDI	31690	Not specified	d polymeric MDI
NDI	> 5000	Rat	
PPDI	≈ 2000	Rat	
CHDI	760	Rat	
HDI	913	Rat	e
HDI	750	Rat	f
HDI	738	Rat	g
TMDI	4180	Rat	
XDI	3200	Rat	h
XDI	840	Mouse	i
HMDI	> 11000	Rat	j
HMDI	9900	Rat	k
IPDI	4825	Rat	l
IPDI	2650	Rat	m
IPDI	2650	Mouse	m
HXDI	1900	Rat	h
HXDI	570	Mouse	i

* For the coding a–m, see p. 39

The oral toxicity figures (LC_{50} rat) of polyisocyanates on the basis of TDI, HDI and IPDI with monomer contents below 0.5 ... 1.0% are higher than 5000 mg/kg. This has been established even with commercial preparations containing solvents [158–163]. Triisocyanates as they are used for the preparation of adhesives (which usually contain approx. 20% dichloromethane), show LD_{50} figures (rat) between 3000 and 4000 mg/kg [171, 172].

There is some data available on the oral toxicity of TDI (2,6-isomer) on two avian species for which LD_{50} figures in the order of 100 mg/kg have been determined [194].

5.1.2 Acute Inhalation Exposure

Exposure time, the physical and chemical nature and concentration of the test material are the most important parameters influencing the outcome of inhalation toxicity studies with laboratory animals. Careful analytical control is essential to assure a constant and homogenous concentration distribution in the inhalation chamber; as pointed out in Section 6, monitoring of airborne isocyanates is a particularly difficult problem because of their specific chemical characteristics.

Figures for the acute inhalation toxicity determined on animals can contribute in assessing the risk to humans after sudden and unexpected exposure to massive concentrations of an airborne chemical. Regulatory bodies often use such figures (together with other toxicological and ecological data) to define the classification of chemical substance within the rules for transportation and handling.

The acute inhalation toxicity is usually described by the LC_{50}. This is the concentration of the airborne test substance at which 50% of test animals (mostly rodents) die during or after an exposure of usually 4 hours. The observation period after the test varies and can be up to 4 weeks. In view of the relativily low vapor pressure of diisocyanates, a concentration of this order can only be achieved at ambient temperature – necessarily the temperature in the exposure chamber – by increasing the concentration above the level of the saturated vapor at this temperature i.e. by exposing the test animals to a mixture of vapor and aerosol. It is obvious that LC_{50} data achieved by aerosol exposure do not allow a fully realistic assessment of the toxic risk. In order one may correlate the LC_{50} data with the maximum vapor levels which can be expected while ambient conditions prevail, the highest possible concentration of diisocyanates at 20 °C are shown in Table 21. It should be noted that the figures are accurate to an order of magnitude only as most of them have been obtained by calculation assuming that the vapors of diisocyanates behave as ideal gases. Concentration figures are specified in mg/m^3 and in parts per million (ppm) which are equivalent to ml/m^3. It is physically correct that both units – mg/m^3 and ppm – may be used for the vapor state whereas an aerosol concentration can be indicated in mg/m^3 only.

5.1.2.1 Diisocyanates

TDI– Acute Inhalation Toxicity
The following figures are communicated in the literature:

LC_{100}: 4380 mg/m^3 6h rat [176]
LC_0: 438 mg/m^3 6h rat [176]
LC_{50}: 102 mg/m^3 ≈ 14 ppm 4h rat [178]
LC_{50}: 350 mg/360 mg/m^3 4h rat [179]
LC_{50}: 58 mg/m^3 ≈ 8 ppm 1h rat [195]
LC_{50}: 480 mg/m^3 1h rat [131]
LC_{50}: 73 mg/m^3 ≈ 10 ppm 4h mouse [178]
LC_{50}: 80 mg/m^3 ≈ 11 ppm 4h rabbit [178]
LC_{50}: 95 mg/m^3 ≈ 13 ppm 4h guinea pig [178]

The wide variation of data is evident. When looking at the details of these studies, particularly when assessing the meaning of the analytical data is in the light of modern isocyanate analysis, one has to conclude that the LC_{50} 4h, rat is about 110 mg/m^3. The LC_{50}, 1h, rat is in the order of 480 mg/m^3. The latter figure is considerably higher than the concentration of the saturated vapor at the exposure temperature.

Table 21. Diisocyanates – concentration of saturated vapor at 20 °C

Product	mg/m^3	ppm
TDI	140	20
Monomeric MDI	0.1	0.01
Polymeric MDI	0.15	0.015
NDI	0.85	0.1
PPDI	72	11
CHDI	34	5
HDI	75	10
TMDI	9	1
XDI	2	0.25
HMDI	11	1
IPDI	4	0.4
HXDI	1.5	0.2

TDI – Other Acute Inhalation Effects

The airway hyperresponsiveness and inflammation induced by exposure to TDI vapor (1 ppm and 2 ppm) were studied on guinea pigs. The animals were challenged at various points of time after exposure with an acetylcholine aerosol. The subsequent increase in airway responsiveness was associated with histologic changes in the trachea and the intrapulmonary airways [196]. A proposed enzymatic mediation of TDI-initiated airway edema was studied on guinea pigs which were exposed to 3 ppm TDI vapor for 1 hour [197].

On male mice the effects of acute exposure at concentrations ranging from 0.05 . . . 14.2 mg/m^3 (0.007 . . . 1.95 ppm) were studied in order to detect the level of sensory irritation caused by TDI vapor (2,4-isomer). This level which was expressed in terms of an RD_{50} (concentration causing a 50% decrease of respiratory rate) was reduced significantly when the concentration and the exposure time were increased. The RD_{50} was for example 1.42 mg/m^3 (0.2 ppm) after a 180 min exposure [198]. The sensory irritation of TDI vapor (2,6-isomer) was studied on male mice at 0.37 . . . 7.6 mg/m^3 (0.05 . . . 1.1 ppm). The RD_{50} 180 after min exposure was determined to be 1.85 mg/m$^3 \approx$ 0.25 ppm [199].

Monomeric and Polymeric MDI

LC_{50}: 370/380 mg/m^3 4h rat, polymeric MDI [179]
LC_{50}: > 400 mg/m^3 2h rat, aerosol (powder) of monomeric MDI [200]
LC_{50}: 490 mg/m^3 4h rat, polymeric MDI [137]

The LC_{50} rat 4h of polymeric MDI is therefore in the order of 450 mg/m^3; this is more than 3 orders of magnitude higher than the saturated vapor at 20 °C.

NDI

LC_{50} data on the basis of an acute single exposure are not reported.

PPDI [142]
LC$_{50}$: 10960 mg/m^3 1h rat (exposure to vapor/aerosol generated by passing air through molten product)
LC$_{50}$: < 500 mg/m^3 4h rat (exposure to particles generated from powdered product)

CHDI [143]
LC$_{50}$: 4540 mg/m^3 1h rat (exposure to vapor/aerosol generated by passing air through molten product)
LC$_{50}$: < 500 mg/m^3 4h rat (exposure to particles generated from powdered product)

HDI
LC$_{50}$: 310–350 mg/m^3 4h rat [179]
LC$_{50}$: 290 mg/m^3 4h rat [144]
LC$_{50}$: 154 mg/m^3 4h rat [201]
LC$_{50}$: 280 mg/m^3 1h rat [201]

For the rat the lowest lethal concentration for a 3 hour exposure period amounted to 60 mg/m^3 (189). Exposure of rats to the saturated vapor for 8 hours did not cause fatalities [188].

TMDI [145]
LC$_{50}$: 700 mg/m^3 4h rat

XDI [190]
LC$_{50}$: 182 mg/m^3 4h rat

HMDI
LC$_{50}$: 295–307 mg/m^3 4h rat [202]
LC$_{50}$: 190 mg/m^3 5h rat [191, 203]

IPDI
LC$_{50}$: 123 mg/m^3 4h rat [148,189]
LC$_{50}$: 670 mg/m^3 4h rat [149]
LC$_{50}$: 260 mg/m^3 1h rat [189]

HXDI [190]
LC$_{50}$: 190 mg/m^3 4h rat

5.1.2.2 Polyisocyanates

Acute Inhalation Toxicity
The acute inhalation toxicity of some polyisocyanates used as components for spray coatings has been tested as the aerosol. They were generated from the solvent containing commercial preparations. The results are as follows [179]:

TDI based adduct type: LC_{50} 4h rat > 3820 mg/m^3
TDI based polyisocyanurate type: LC_{50} 4h rat > 2462 mg/m^3
TDI/HDI based polyisocyanurate type: LC_{50} 4h rat > 3003 mg/m^3
HDI based adduct type: LC_{50} 4h rat 500 mg/m^3
HDI based biuret type: LC_{50} 4h rat 400–425 mg/m^3
IPDI based adduct type: LC_{50} 4h rat > 5362 mg/m^3

In the same study the LC_{50} figures of a number of polyols (used as reaction partners of the aforementioned polyisocyanates and of several non-isocyanate coating systems) were evaluated. In all cases an LC_{50}, 4h, rat, above 2000 mg/m^3 was the result confirming that the acute inhalation toxicity of TDI and IPDI based polyisocyanates is of the same magnitude as polyols and other coating materials. The HDI based types show a somewhat higher toxicity; for the HDI based polyisocyanurate type an LC_{50} 4h rat of 430 . . . 450 mg/m^3 has been determined [163].

Other Acute Inhalation Effects
A comparison of HDI with its biuret derivate in an inhalation study with mice (the biuret was administered as aerosol) has shown that the biuret produced primarily pulmonary irritation whereas the HDI was a potent upper respiratory tract irritant [204]. The biuret type was studied as aerosol at concentrations of 25 . . . 131 mg/m^3 on male mice; the exposure time was 3 hours during which the respiratory rate of the animals was recorded. The product acted primarily as a pulmonary irritant; an RD_{50} of 57.1 mg/m^3 was determined [204]. Pulmonary oedema evident in guinea pigs after 4 h inhalation of an HDI-polyisocyanurate derivative aerosol at concentrations of 2.9 . . . 39 mg/m^3 can be detected and evaluated by a single tracer technique [205].

5.1.3 Effects of Repeated Subacute Inhalation Exposure

Numerous studies have been conducted (specifically with TDI, MDI and HDI) exposing animals repeatedly to non lethal concentrations of the test substances. The aim was to ascertain

– the effect of the chemicals on the respiratory organs and to find out whether it was permanent or reversible and whether other organs were affected,
– the mechanisms which control the observed sensitization phenomena,
– whether a possibility exists to diagnose sensitization on humans by an immunological test procedure.

Since these problems are rather complex, the studies are ongoing. Only a survey on the actual state of affairs so far can therefore be given.

5.1.3.1 TDI

It was shown many years ago that a subacute exposure (4 . . . 6 times on different days for 4 . . . 6 hours daily) to concentrations up to 3 . . . 5 ppm (approximately 20 . . . 35 mg/m^3) does not kill rodents (176, 177). If the administration of TDI vapor is extended to, for example, 38 exposures (6 h/day, 5 d/week), rats, rabbits and guinea pigs show chronic irritation of the tracheal and bronchial mucosa even at 1.0 ppm (approximately 0.7 mg/m^3) TDI [206]. TDI was compared with a number of other sensory irritants by exposing mice to its RD_{50} concentration (2.9 mg/m^3 approximately 0.4 ppm) for 6 h/day over 5 days. The chemical, like all other tested compounds, caused lesions in the nasal cavity with a distinct anterior–posterior severity [207]. An inhalation study on rats was performed at levels of 3.5, 7.1, 35.6, 71.2 mg/m^3 (0.5, 1, 5 and 10 ppm) respectively. On consecutive days (6d/w) 24 exposure periods of 6h at the 3.5 mg/m^3, 10 at the 7.1 mg/m^3 and 2 . . . 5 at the 71.2 mg/m^3 levels were carried out. The high levels were lethal for most animals. Death was due to mechanical blocking of the respiratory passages by mucosal tissue detached from bronchi and trachea. At the two lower levels fatality was between 45 and 75% for younger animals (initial body weight 91 . . . 124 g). After 24 exposure periods at 3.5 mg/m^3 (with an interval of 4 weeks after the 12th exposure) no fatality occurred among animals with an initial body weight of 140 . . . 180 g. Of the surviving animals in the 7.1 mg/m^3 group, the pulmonary changes which were observed reverted partially; complete remission took place in the 3.5 mg/m^3 group [177]. In a study with male dogs the average TDI concentration was 10.7 mg/m^3 (approximately 1.5 ppm). They were exposed over a period of 4 months 35 . . . 37 times; the duration of each exposure was 30 . . . 120 min. The animals showed lachrymation, coughing, restlessness and expectoration of a white frothy material. The autopsy indicated mild congestion and inflammation of trachea and large bronchi [176].

Exposure of guinea pigs on 3 occasions to 0.01 . . . 5.0 ppm TDI resulted in respiratory sensitization; they tested positive after skin treatment with TDI [208]. Pulmonary hypersensitivity and sensitization of guinea pigs occurred after exposure to 0.25 ppm (approximately 1.8 mg/m^3) 3h/day for 5 days. 3 out of 16 animals exposed developed antibodies when challenged with TDI guinea pig serum albumin antigen. Consistent increase of pulmonary hypersensitivity was not observed [101]. The findings have been confirmed by other authors [209]. On guinea pigs which were exposed to TDI vapor of various concentrations between 0.85 and 71.2 mg/m^3 (0.12 . . . 9.8 ppm) 3 h/day for 5 days assays for TDI specific antibodies, skin and pulmonary sensitivity were performed. No response was found on animals exposed to 0.85 mg/m^3 whereas 55% of the animals displayed TDI specific antibodies when exposed to $\geqslant$ 2.56 mg/m^3 (dose related). Pulmonary sensitivity was observed when challenged with TDI protein antigen at concentrations exceeding 2.6 mg/m^3; levels higher than 14.2 mg/m^3 resulted in pneumotoxicity and limited pulmonary hypersensitivity reaction [210]. Induction of respiratory and dermal sensitivity on guinea pigs was achieved by sensitizing them through exposure to 6.7 mg/m^3 (approximately 0.92 ppm) for 3 h/day on 5 consecutive days. 4 out of 8 animals developed dermal sensitivity by day 8. Inhalation challenge by 0.4 mg/m^3 TDI was negative but challenge by a TDI guinea pig serum albumin adduct produced immediate onset responses in 4 of 8 animals [211]. Guinea pigs after repeated exposure to 1.4 ppm TDI (3h/day on 5 consecutive days) developed bronchitis; this treatment was followed by challenge with air containing 10% carbon dioxide. The ventilatory response was greatly diminished; it recovered during the next 40 days. Challenge after exposure to 0.02 ppm TDI did not influence the ventilatory response [212].

5.1.3.2 MDI, NDI and HMDI

The exposure of rats to MDI for 0.5h/day, 5 days per week for 2 weeks to concentration levels of 2, 6.5 and 28.5 mg/m^3 produced no product related effects [128]. In a sensitization study guinea pigs were exposed to 0.94 mg/m^3 (monomeric MDI) and 4.63 mg/m^3 (polymeric MDI) respectively for 4h/day on 5 consecutive days. Afterwards the animals were challenged with 2.5 mg/m^3 (monomeric) 4.6 mg/m^3 (polymeric), respectively. The polymeric MDI produced a slight but nevertheless significant elevation of airway resistance; monomeric MDI produced only a slight tracheobronchitis [213]. Polymeric MDI was the subject of a subacute inhalation study in which rats were exposed to aerosol concentrations of 2, 5 and 15 mg/m^3 for 2 weeks. No significant effects were found [137, 214]. The pulmonary irritation potential of MDI and HMDI was studied by exposing mice to aerosols of these 2 chemicals with a concentration varying from 17 . . . 67 mg/m^3. The time of exposure was 240 minutes; respiratory pattern and frequency were recorded during exposure. Unlike other isocyanates tested with this animal model, MDI and HMDI acted primarily as pulmonary irritants evoking only very little sensory irritation. An RD$_{50}$ of 32 mg/m^3 was determined for MDI. For HMDI an RD$_{50}$ figure of 40 mg/m^3 was determined [215].

Repeated exposure of rats to an NDI aerosol for 6h per day on 5 subsequent days (28 days observation) resulted in an LC$_{50}$ of 9.1 mg/m^3; for mice which were treated under the same conditions an LC$_{50}$ of 10.7 mg/m^3 was found [216].

5.1.3.3 HDI compared with TDI

Rats and mice were exposed to 1.2 mg/m^3 (approximately 8.4 ppm) HDI for 4h/day during 40 days. Acute reaction was characterized by respiratory tract irritation, loss of body weight and hematologic disorders. The autopsy revealed lung lesions and damage of the upper respiratory tract [217]. After exposure for 6h/day on 5 consecutive days the following LC$_{50}$ figures were found: 1.5 ppm on mouse, 2.9 ppm on rat, 1.8 ppm on guinea pig; for comparison: TDI studied under the same conditions showed the following figures: 1.4, 3.5 and 2.0 ppm [218]. Inhalation tests with HDI on rats, mice and guinea pigs at 0.1 ppm for 6h/day, 5d/week for 4 weeks exhibited a minor effect on weight increase; at 0.03 ppm effects were not detected; the results on TDI were comparable [219]. On male rats an LC_{50} of 4.2 ppm after exposure of 4h/day for 5 consecutive days was determined [201].

5.1.4 Dermal Toxicity – Irritation and Sensitization

The risk of acute intoxication following skin contact with polyfunctional isocyanates is rather low. The LD_{50} figures of various diisocyanates usually determined on rabbit confirm that the amount which penetrates the skin should be extremely small and not capable of causing systemic toxic syndroms. These substances are, on the other hand, active irritants of the skin and mucous membranes leading to erythema and, in extreme cases, to corrosion. Test species for determining the extent of skin damage are usually rabbit and guinea pig; the latter test animal is specifically suitable for assessing the skin sensitization potential which has been found for several diisocyanates.

Published data on animal studies characterizing the acute dermal toxicity and the irritation capacity of diisocyanates are manifold; they have been obtained under various test conditions; the results are often not completely uniform, sometimes also contradictory. This becomes evident from data which are communicated for the various products in the following subsections.

The rating presented in Table 22 for various diisocyanates is based on studies performed according to guidelines which were formulated by experts of OECD countries and adopted subsequently [220].

Working Groups in the United States and in Japan have performed detailed animal studies into the skin sensitization potential, particularly of TDI, MDI and HMDI.

5.1.4.1 TDI

Amounts up to 16 000 mg/kg produced severe local irritation but failed to kill rabbits or to produce injury to their internal organs [176]. 10 000 mg/kg however were determined as the LD_{50} for rabbits [195].

Table 22. Diisocyanates – irritant effect on skin and mucous membranes

Product	Effect on	
	Skin	Mucous Membranes
TDI 80	Corrosive	Strongly corrosive
MDI polymeric	Slightly irritative	Non irritative
HDI	Corrosive	Corrosive
IPDI	Strongly irritative/Corrosive	Strongly corrosive
HMDI	Moderately irritative	Non irritative

TDI was compared with MDI using the Draize method for dermal effects on rabbits. TDI was more corrosive than MDI on rabbit skin and subsequently classified as a "medium irritant" [221].

Only rarely was sensitization after contact with TDI observed in guinea pigs [222]. It was found to severely irritate ocular mucosa whereas MDI produced only slight irritation. Application of a TDI solution in olive oil to the skin of the same species induced contact sensitivity after one week; resulting respiratory tract sensitivity was confirmed after 2 weeks by bronchial challenge with TDI and TDI-protein conjugates and by serological tests [223]. The severity of dermal sensitization of guinea pigs depends on the dose used for induction and challenge [224]. In a dose-response study with several specific species, the animals were sensitized by applying a solution of TDI in ethylacetate to the skin of the animals; they were challenged later by treating the skin again with the TDI solution. Sensitization was confirmed by the resulting swelling of the ear; its extent depended on the TDI concentration in the test solution and reduced in time [225, 226, 227]. The skin sensitizing ability of *ortho*-, *meta*- and *para*-tolyl isocyanate is comparable to TDI, although weaker [228].

Allergic dermatitis developed in mice sensitized dermally to TDI by inhalation exposure to the test reagent [229]. Development of contact sensitivity in guinea pigs after inhalation of TDI vapor is suggested [230]. The coexistence of both dermal and respiratory sensitivity after repeated exposure to airborne TDI has been determined in guinea pigs. Inhalation challenge resulted in dermatitis [211].

5.1.4.2 MDI –Monomeric and Polymeric

10 000 mg/kg applied to the skin of rabbits caused death [231]. MDI was considered a severe eye irritant but it was less corrosive on rabbit skin than TDI according to the Draize method. On the skin of the guinea pig it induced cutaneous sensitivity similar to a contact allergy [221]. Only mild irritation was observed when polymeric MDI was applied to the skin of rabbits; the effects cleared after 5 days; no gross pathological effects were evident 8 days after testing [128]. MDI is a mild to moderate irritant to the skin and eyes of animals [138].

Contact sensitivity was induced by epidermal application of an MDI solution in ethylacetate to the back of a specific species of mouse. The treatment resulted in marked ear swelling; the extent of swelling was comparable to that induced by TDI [232]. Monomeric and polymeric MDI applied to the skin of guinea pigs which were previously exposed to MDI by inhalation induced a systemic sensitization [213].

5.1.4.3 Other Diisocyanates

NDI
No data has been ascertained on the dermal effect of NDI.

PPDI
The substance is reported to be mildly irritating for rabbit skin and eye; it induces mild sensitization in guinea pigs [142].

CHDI
The substance is reported to be mildly irritating for rabbit skin and irritating for eye; it induces mild sensitization in guinea pigs [143].

HDI
The dermal LD_{50} for rabbit has been determined as 593 mg/kg [188]. On the basis of symptoms observed in guinea pigs after application of an HDI solution to the skin, it has been suggested that allergic dermatitis may occur in workers exposed to the substance [233]. The substance, when tested on mice, was proven to be nearly as irritating to the skin and mucous membranes as TDI [189].

TMDI
A dermal LD_{50} in rat of > 7000 mg/kg, a strong irritation potential to the skin and a mild potential for mucous membranes have been reported [145].

XDI
Moderate to severe irritation resulted after application to the skin and severe irritation when administered to the eyes of rabbits. The LD_{50}, dermal, rabbit has been determined as > 5 g/kg [190].

HMDI
The LD_{50}, dermal has been determined on rabbit as $> 10\,000$ mg/kg. Skin irritation followed by sensitization was induced in the guinea pig. At the eye of the rabbit only mild and reversible irritation occurred [147].

Several clinical findings on humans (see Section 7.6) initiated a series of investigations on laboratory animals. Extensive sensitization of guinea pigs was observed after topical as well as intradermal and intraperitoneal administration of HMDI [234]. The skin sensitization potential was studied more in depth on topical exposure of guinea pigs and a specific mouse species. The guinea pigs were challenged 7 days later by patch-testing; the extent of erythema thus formed was clearly dependent on the administered dose. Mice were challenged by an ear-swelling assay technique; monitoring the ear thickness change indicated a clear dose-response relationship. For mice an SD_{50} (dose needed to sensitize 50% of the animals) of 0.2 mg/kg was calculated [235]. In a later study the dermal sensitizing potencies of HMDI, TDI, MDI and HDI were evaluated and compared using the mouse ear-swelling test. An SD_{50} value of 0.24 mg/kg resulted for HMDI (this being in good agreement with the aforementioned figure) compared to 5.3 mg/kg for TDI, 0.73 mg/kg for MDI and 0.088 mg/kg for HDI [184].

IPDI
The LD_{50}, dermal has been determined on rat as 530–1060 mg/kg after 4 h exposure [193]. An LD_{50}, dermal on rat of > 7000 mg/kg, a strong irritation

potential for the skin, but no irritation of the mucous membranes are also reported [149]. The substance is, according to [236], extremely irritating to the skin and mucous membranes.

HXDI

Moderate irritation is reported for the skin and severe irritation for the eye of the rabbit; the LD_{50} was ascertained as > 5 g/kg [190].

5.1.5 Special Animal Studies

Various approaches are aiming to generate information on the circumstances of diisocyanate caused respiratory disease. They are briefly dealt with in this paragraph; the interested reader may refer to the original literature.

5.1.5.1 Cholinesterase Inhibition

A supposed inhibition of guinea pig cholinesterase after inhalation of diisocyanates has not been confirmed in a study with HDI on guinea pigs which were exposed to 0.5, 1.8 and 4.0 ppm for 6 hours [237]. In contrast to this finding are the results of in vitro studies with human cholinesterase and 2,4- and 2,6-TDI, HDI and several mono-IC's. Inhibition effects of various intensities have been observed [57]. The isocyanate is linked to the active site of the enzyme; the inhibition is reversible [238]. A study performed on TDI, HDI, phenyl-IC and cyclohexyl-IC by another researcher has confirmed the findings [58]. The inhibition potency of MDI has been compared with TDI and HDI: it is in the order of HDI > TDI > MDI [239]. A dosimeter on the basis of the cholinesterase inhibition reaction allows the detection of minute concentrations of HDI and TDI in air [240]. The response of the guinea pig tracheal smooth muscle to TDI vapor after repeated inhalation at 0.02 . . . 0.03 ppm (5h/day for up to 20 days) was the subject of a series of studies: the observed increase in maximal tension and the shift of the dose-effect curve compared to the controls suggested a direct influence of TDI on this muscle [241, 242, 243].

5.1.5.2 Immunological Studies

The focal points of animal studies have been and are still immunological and pharmacological approaches to evaluate the mechanism of sensitization by diisocyanates; the question which of these two biochemical pathways – possibly simultaneously – is responsible for the onset of an asthmatic reaction after challenging a sensitized animal is of utmost importance for the diagnosis and therapy of isocyanate sensitization in humans. The problem of immediate and/or delayed onset of an asthmatic reaction after isocyanate exposure which has been observed in sensitized humans is the subject of studies with animal models [244, 245, 246]. The pharmacological action of TDI has been evaluated in vitro on human blood leucocytes [247].

Conjugates containing mono-IC's as haptens are more suitable as antigens for the detection of hapten specific antibodies in the blood serum of isocyanate sensitized guinea pigs than those on the basis of diisocyanates. This is due to cross linking, and other side reactions, of the bifunctional product which forms a conjugate with a suitable protein. The presence of TDI antibodies in guinea pigs sensitized to TDI has been identified by an antigen formed from p-tolyl mono-IC and ovalbumin [248, 249]. A hexyl-IC-ovalbumin conjugate was studied as antigen to detect HDI antibodies in guinea pigs sensitized to HDI by the inhalation route [38] and by intradermal injection [250]. Nasal allergy was induced on guinea pigs by painting a 10% TDI solution in ethylacetate on the nasal cavities. Provocation with a 5% solution which was performed three weeks later initiated exertional breathing and prolongation of the expiratory phase. By repeating the challenge with reduced amounts of TDI the severity of the attacks was not notably reduced [251].

Induction of TDI antibodies in mice after intraperitoneal administration of the chemical has been confirmed by a positive in vitro reaction of the blood serum with a TDI serum-albumine conjugate [252]. The immunization of guinea pigs has also been studied by parenteral administration of well character- ized conjugates like TDI- and HDI-human serum albumin. Cross reactivity of the resulting antibodies with antigens formed by these diisocyanates after conjugation with proteins was confirmed in vivo [253], see also [254]. Intra- venous challenge with these conjugates resulted in distinct and significant pulmonary responses [255]. A suspension of MDI was given intrabronchially to dogs. The animals showed a systemic immune response to an MDI dog-serum albumin conjugate [256].

Immunization of dogs by repeated exposure to 0.02 ppm TDI induced the development of systemic immune responses to TDI dog-serum albumin [257].

A significant release of histamine from leucocytes of the nasal mucosa was noted when stimulated in vitro by a TDI guinea pig albumin conjugate [258]. The in vitro histamine release from leucocytes induced by a TDI human serum-albumine conjugate was considerably higher in guinea pigs which had been exposed to approximately 1 ppm TDI compared to non exposed animals [259]. The nature of airway mucosal disease and the histaminic reactivity of guinea pigs after exposure to 3 ppm TDI for 4 h/day on 5 consecutive days was assessed at various time intervals. Bronchial reactivity was determined by control of the airway conductance as a function of increasing doses of histamine aerosol [260].

Studies on respiratory effects of isocyanates and the sensitization of laborat- ory animals via the dermal and inhalation route are the subject of several review publications [183, 261, 262].

5.1.6 Mutagenicity

A number of in vitro and in vivo mutagenicity tests have been carried out by several researchers. As such studies have to be conducted in an aqueous environment the results do not definitely indicate whether they relate to the test substance or to its reaction products with water. This has to be taken into account while interpreting isocyanate mutagenicity data.

5.1.6.1 TDI

There are conflicting reports about the mutagenicity of this compound. According to [263, 264] a test with Salmonella Typhimurium proved negative whereas a later study showed a dose dependent mutagenic response in some strains of these bacteria by activation with liver microsomes. An adduct type polyisocyanate on the basis of TDI however proved negative [265]. The conclusions drawn by the authors were the subject of critical discussion [266, 267]. Since the investigators used dimethylsulfoxide to facilitate the dissolving in the aqueous bacteria preparation, the results must also be seen in light of a study [268] discussed in detail in Section 5.2.1. As the use of water miscible solvents furthers the formation of diamines from diisocyanates, the positive mutagenicity of TDI is not surprising because toluene diamine (TDA) is a potent mutagen.

The individual TDI-isomers were also found to be mutagenic in two Salmonella Typhimurium strains with enzymatic activation but not in two other strains [269].

Two in vitro cell transformation assays with human and hamster kidney cells showed a negative result [270] as did an in vivo micronucleus test performed on bone marrow erythrocytes from rats and mice which were exposed to 0.35 and 1.05 g/m^3 TDI for 6 h/day on 5 days/week over 4 weeks [271].

TDI induced chromosome aberrations in human blood lymphocyte cultures in the absence as well as in the presence of metabolic activation [272].

5.1.6.2 MDI

In three independent studies, mutagenic effects were observed in two equal Salmonella strains in the presence of liver microsomes [128, 265, 273]. On all other strains tested no mutagenicity was observed with or without activation [128]. Positive findings are also reported in [274]. Chromosome aberrations and sister chromatid exchanges with and without metabolic treatment were induced by polymeric MDI in human blood lymphocyte cultures [272]. In an in vivo mouse micronucleus test MDI produced no increased incidence of micronuclei formation [137].

5.1.6.3 NDI, TMDI and IPDI

No mutagenicity data have been reported.

5.1.6.4 PPDI and CHDI

The substances are non-mutagenic in the Salmonella Typhimurium test [142, 143].

5.1.6.5 HDI

The chemical is non-mutagenic in the *Salmonella Typhimurium* test [268].

5.1.6.6 XDI and HXDI

The chemicals gave a negative bacterial mutation test [190].

5.1.6.7 HMDI

The compound is reported to test negative in the *Salmonella* test [202].

5.1.7 Carcinogenicity

5.1.7.1 TDI

A carcinogenicity study with groups of male and female rats and mice was performed by exposing the animals to 0 (control), 0.37 and 1.1 mg/m^3 (approximately 0.05 and 0.15 ppm) vapor of commercial TDI (isomer ratio 80:20) for 6 h/day, 5 days/week for approximately 2 years. These concentrations were based on the results of a preceeding range finding study in which groups of rats, mice and hamsters were exposed to TDI vapor (0 as control, 0.73 and 2.2 mg/m^3, approximately 0.1 and 0.3 ppm) for 6 h/day, 5 days/week for a total of 20 . . . 22 exposures.

The results of the long term studies are the subject of several publications [182, 271, 275]. Type and incidence of tumors and the number of tumor-bearing animals of either species did not reveal any carcinogenic effect. Hematology and biochemistry parameters and urinanalysis did not show any exposure related influence. Signs of toxicity indicated that 0.15 ppm as highest exposure level was close to the maximum tolerable concentration. In the rat study, the animals of both sexes gained at this concentration less weight compared to the other dose levels during the first 12 weeks. In the mouse study, a dose related increase in the incidence of chronic rhinitis was observed.

The aforementioned study was designed to simulate the exposure situations arising in the workplace. Its outcome is subsequently useful in assessing the risk of chronic exposure of humans, including carcinogenicity.

The oral administration of TDI used in another long-term study [276] does not reflect this exposure scenario. Groups of rats and mice (both sexes) were dosed with a TDI-corn oil mixture by gavage 5 days/week for 105 . . . 106 weeks. The daily doses were 60 or 120 mg/kg of body weight for female rats and female mice; 30 or 60 mg/kg were administered to male rats and 120 or 240 mg/kg to male mice. The doses were based on subchronic studies which showed up effects on body weight and bronchopneumonia with doses exceeding 120 mg/kg.

Under these conditions, survival time for dosed rats was shorter than that of controls and body weight gain was decreased in exposed animals relative to controls. A dose-dependent cumulative toxicity resulting in excessive mortality indicated that the maximum tolerated dose had been exceeded for rats. Survival was also compromised in that high dose male mice had a shorter lifespan than controls. Another problem observed in the study was the instability of the mixture of TDI and corn oil. Analytical evaluations have confirmed that the two products formed an undefined reaction product which was also administered to the animals. Subsequently, actual gavage concentrations were reported as 77 . . . 90% of theoretical values. It can also be taken as certain that TDI is hydrolysed to toluene diamine (TDA) under the acidic conditions of the stomach. The 2,4-TDA is an animal carcinogen [129, 277, 278].

The conclusion of the study was that subcutaneous tumors in male and female rats, liver, pancreatic and mammary gland tumors in female rats as well as liver and vascular tumors in female mice were related to the oral administration of TDI.

The results of the inhalation exposure and the gavage application studies were discussed by the Working Group on the Evaluation of the Carcinogenic Risk of Chemicals to Humans of the International Agency for the Research on Cancer, which is a body of the World Health Organization, in June 1985. The Working Group concluded that, "there is sufficient evidence for the carcinogenicity of toluene diisocyanate to experimental animals" and "there is inadequate evidence for the carcinogenicity of toluene diisocyanate to humans" [129].

5.1.7.2 Other Polyfunctional Isocyanates

Results of long-term toxicity studies on compounds other than TDI have not been published.

5.2 Ecological Aspects – Ecotoxicity

The ecological behavior of polyfunctional isocyanates i.e. their interaction with water, soil and air and their influence on living organisms in these environments are mainly determined by their reactivity, in particular with water. As these products are practically insoluble in water, the mixing conditions for studies in an aquatic environment have to be precisely defined. It is obvious that they should be adapted as close as possible to scenarios which may occur in practice. The polyureas which are formed as reaction products are solid, insoluble and practically inert substances; this has to be considered when the influence of these isocyanates on aquatic animals is investigated. The concentration of diisocyanate vapor on escape into the air is usually very low compared to the air's average water vapor content which would usually be the only reaction partner. The kinetics of gas phase reactions have to be considered if the fate of the airborne chemical must be assessed. Numerous possibilities for interaction exist however between a polyfunctional isocyanate if it should accidentally ooze into the soil. If it is humid, polyurea will in this case be the main reaction product.

TDI (and to a minor extent MDI) have been studied specifically in this respect. The state of present knowledge is the subject of several reviews [279, 280, 281]. The ecological behavior of TDI has also been included in a research programme on industrial chemicals sponsored by the US government [282].

5.2.1 Reaction with Water

The consequences of a TDI spillage into stagnant and running water have been studied in several tests at pH conditions simulating a slightly acidic or slightly alkaline situation. Polyurea was the main reaction product. The amount of toluene diamine (TDA) generated under the test conditions was $10^{-6}\%$ in comparison with the original TDI [280].

The hydrolysis of TDI and HDI if tested after dissolving in a water miscible solvent takes a different course [268]: solutions of the two products in acetone were mixed with water at ambient temperature. The reaction progression of the diisocyanates with water and the formation of diamines were followed by photometric analysis; polyurea, which was assumed to be insoluble in water, was determined gravimetrically. The outcome of the study is presented in Table 23. The higher reactivity of TDI versus HDI is obviously responsible for the larger yield of polyurea in the test with TDI.

The same authors studied the concentration decrease rate of TDI and HDI in water at two temperature levels using acetone as solution mediator. The results are shown in Table 24. The data in Table 25 indicate that the oxidation of TDA and HDA when tested in tap water and in river water proceeds rather

Table 23. TDI and HDI – Products of hydrolysis

TDI (139 mg), HDI (136 mg) dissolved in each case in 25 ml acetone, solution mixed with 1000 ml water, result after completion of reaction (percent figures relate to the amount of diisocyanate applied)

TDI:
56% hydrolysed to toluene diamine (TDA)
36% reacted with TDA to polyurea
72% consequently deposited as polyurea in the aqueous phase
21% remaining as TDA in aqueous solution
92% had reacted in total

HDI:
77% hydrolysed to hexamethylene diamine (HDA)
13% reacted with HDA to polyurea
26% consequently deposited as polyurea in the aqueous phase
64% remaining as HDA in aqueous solution
90% had reacted in total

Table 24. TDI and HDI – hydrolysis study: TDI (20 . . . 30 mg) and
HDI (200 mg and 2 mg) dissolved in each case in 25 ml acetone, mixed
with 1000 ml water; concentration decrease after various time intervals

20 . . . 22 °C:		
	200 mg HDI:	50% after 5 min
		90% after 30 min
	2 mg HDI:	50% after 10 min
		90% after 50 min
20 . . . 30 mg TDI:		90% after 5 min
4 . . . 6 °C:		
	200 mg HDI:	50% after 20 min
		90% after 120 min

Table 25. Oxidation of TDA and HDA in aqueous solution: concentration of the diamines: 10 . . . 20 mg/l, concentration decrease after various time intervals:

Tap water, 20 . . . 23 °C:	
TDA: 30% after 30 days	
HDA: 54% after 30 days	
Tap water, 4 . . . 6 °C:	
TDA and HDA: 7 . . . 10% after 30 days	
River water, 20 °C:	
TDA: 20% after 25 days	
HDA: 50% after 10 days	
River water, 4 . . . 6 °C:	
TDA: 35% after 30 days	
HDA: 54% after 30 days	

slowly. The higher rates determined in river water as compared to tap water may be caused by the presence of bacterial support of biochemical oxidation.

With regard to the behavior of TDI in wet soil, several model experiments have shown that the chemical is converted, after blending with wet sand, to polyurea at a rapidly decreasing rate. The finding that after 8 days 3.5% of TDI remained unreacted can be explained by the encapsulation within a crust of polyurea which delays the further entrance of water [281].

A case history of a large TDI spillage into a swampy, wet, forest soil which occurred in The Federal Republic of Germany in 1975 showed that one year after the accident neither TDI nor TDA could be found in soil samples [280]. Thus the biological stability of TDI-based polyurea became obvious. A degradation study of polyureas obtained from ^{14}C-labelled TDI and MDI in different agricultural soil revealed that no aromatic amines were formed nor was radioactive carbon dioxide liberated [283].

5.2.2 Effect on Aquatic Organisms

The influence of TDI and MDI on aquatic animals and on bacteria in water has been the subject of several studies. Freshwater Minnows were exposed, under still conditions to TDI in rather high concentrations in order to determine the LC_{50} after 24 . . . 48 h. The results of these tests as well as determinations of the toxicity of TDI on Grass Shrimp must be interpreted with reservation. Researchers noticed that most mortalities occurred during the first 12 hours of the test. A concurrent decrease of the pH was observed as a result of the generation of carbon dioxide by the TDI-water reaction. This could have contributed to the toxicity of TDI toward these aquatic species [284]. A rather high toxicity was shown for the copepod [285].

Other studies have confirmed that in view of the distribution problems and the reactivity of TDI the results are inconsistent but the following can be concluded [281]:

1. TDI and MDI are not appreciably toxic to bacteria;
2. TDI and MDI are not appreciably toxic to daphnia if dispersed in water with moderate efficiency; no negative effects on their reproduction were observed;
3. Results on fish toxicity of TDI and MDI were rather inconsistent as harmful effects due to oral ingestion or mechanical violation of body tissues could not be excluded.

The broad finding was that immediate toxic effects of TDI and MDI are rather low.

It is worth mentioning that the Guidelines 202, 203, 209 and 302 adopted by the OECD for testing the effects of chemicals in an aqueous environment do not give advice on how substances which are insoluble in and reactive with water should be handled.

5.2.3 Fate of Airborne TDI

Many investigations have been carried out on the fate of airborne TDI. As outlined earlier in this paper, TDI is predominantly used for the production of flexible foam slabstock and mouldings. Exhaust facilities assure that the concentration of TDI vapor in the workplace does not exceed the limit in line with occupational health experience and requirements (see Section 6). The TDI stack concentration has been determined as up to 10 mg/m^3 (approximately 1.4 ppm); this represents about 0.005% of the total TDI used [286]. Other controls have shown levels between 0.1 and 17.7 mg/m^3 (approximately 0.01 . . . 2.4 ppm) [287]. The exhaust air 2,6-isomer content is higher than in the 80: 20 TDI starting material; this is because of the higher reactivity of the 2,4-isomer with the components in the foam process [288]. As concentrations at the fenceline of polyurethane foam plants must not exceed 0.003 mg/m^3 (approximately 0.0004 ppm) TDI in some countries (and lower limits in others) processes for emission control using activated carbon scrubbing have been studied. The aim is to adsorb TDI and chlorofluorocarbon blowing agents simultaneously and recycle the latter subsequently [289, 290].

Several researchers have investigated the kinetics of the reaction of TDI vapor in the atmosphere and the reaction products [291–294]. The results are rather inconsistent; this is, amongst other things, due to the limited volume of the reaction chambers (several cubicentimeters up to some liters) and the not always fully efficient analytical procedures. No aromatic diamines were detected.

A study in a 17 m^3 polytetrafluoroethylene lined chamber was carried out. Parameters such as TDI concentration and air humidity were varied, the influence of known air pollutants like ozone, OH-radicals, urban hydrocarbon mixtures and ammonium sulphate particles as well as the effect of triethylene diamine (TEDA)–(a commonly used catalyst which is co-emitted with TDI) were included in the experiments which were performed in darkness and under irradiation. An important aim of the study was to reveal whether TDI would be converted to toluenediamine (TDA) under the conditions of the study.

Table 26 gives a survey of the TDI removal rates. On the basis of the experimental conditions, the following results were obtained:

– The TDI loss rate in air of varying humidities in darkness was low (approx. 15%/h;
– Irradiation did cause an increase in loss-rate which was mainly attributable to the influence of free radicals and, to some extent, to air pollutants present;
– The addition of TEDA (triethylenediamine) considerably increased the loss-rate;
– It became obvious that adsorption on the surface of the reaction chamber was a significant removal mechanism.

Table 26. TDI removal rates

Experiment	Urban mix	Irradiation	TEDA	Other species	Average removal rate*	Net loss rate*
1	–	–	–	–	0.15	0
2	–	+	–	–	0.36	0.21
3	+	+	–	–	0.36	0.21
4	+	+	–	5 ppm Ammonia	0.33	0.18
5	+	+	2 ppm	–	0.99	0.84
6	+	+	–	0.1 mg/m^3 Ammonium Sulphate	0.40	0.25
7	–	–	–	–	0.35	0
8	–	+	–	4 ppm Nitrous Oxide	0.38	0.03
9	+	–	0.2 ppm	–	0.36	0.01
10	+	+	0.2 ppm	–	0.55	0.20

* The unit for the removal rate is hr^{-1}; e.g. 0.15 hr^{-1} is 15%/hr. The net loss-rate is the TDI loss rate minus wall loss-rate

A very important result is that no TDA was found above the detection limit of 10 ng/ml. This corresponds to a maximum conversion of 0.05% TDI to TDA [295,296]. This finding underlines the critique to a report according to which appreciable amounts of TDA had been formed from TDI in the atmosphere of a foam plant [297]. The data communicated by the authors were interpreted as false readings due to interference in the analyses [298]; this statement was accepted [299].

In view of the extremely low volatility of MDI there is no particular concern about emission problems. HDI, as one of the more volatile products, does not pose specific emission problems either because it is applied in practice as polyisocyanate containing only very small amounts of the monomer.

6 Occupational Hygiene

The occupational hygiene and health aspects of monoisocyanates were discussed in Section 3. It has become evident that detailed experience of the health hazards only exists for methylisocyanate whereas the respective information on other monoisocyanates is rather limited. There is little doubt that their mechanism of action is similar but the wide spectrum of vapor pressures of the individual products has to be considered when assessing the hazard. The hazards of poly-functional isocyanates which may threaten humans during handling and process-ing require therefore that stringent occupational hygiene measures are taken: this relates specifically to compliance with the maximum admissible concentration of isocyanates in the atmosphere in the workplace for which regulations or at least recommendations have been established by many countries.

Numerous publications particularly in medical journals—an overview will be given in Section 7—deal with syndroms and diseases attributable to di- and polyisocyanate exposure. The majority of cases are caused by TDI and MDI; most of them occurred in plants where these products are processed. Recom-mendations for occupational hygiene are mainly based on observations and experiences gained from those people directly involved with the chemical.

It goes without saying that all technical equipment needed for processing di- and polyisocyanates has to be in accord with safety requirements. It is also self-evident that employees should wear suitable working suits, gloves and safety goggles when handling these products. Detailed recommendations are given in the technical information bulletins from isocyanate suppliers (e.g. [131]).

The problem of ascertaining the maximum concentration of airborne isocyanate which represents no specific hazard to employees is rather complex and needs detailed discussion.

6.1 Definition of Exposure Limits

The maximum admissible concentration of a dangerous product in the work-place atmosphere has been defined by expert groups in several countries in a somewhat different manner. This is illustrated in the following by the situation in The Federal Republic of Germany, The United States of America, France and The United Kingdom.

6.1.1 Federal Republic of Germany

The German 'Senatskommission zur Prüfung gesundheitsschädlicher Arbeitsstoffe' defines a 'MAK-Wert' (Maximale Arbeitsplatz-Konzentration) and a 'Kurzzeitwert'-(Short-Time-Value) concept [300]:

"The MAK value (maximum concentration value in the workplace) is defined as the maximum permissible concentration of a chemical compound present in the air within a working area (as gas, vapor, particulate matter) which, according to current knowledge, generally does not impair the health of the employee nor cause undue annoyance. Under these conditions, exposure can be repeated and of long duration over a daily period of eight hours, constituting an average work week of 40 hours (42 hours per week as averaged over four successive weeks for firms having four work shifts). As a rule, the MAK-Value is integrated as an average concentration over periods of up to one workday or one shift. When establishing the MAK values, primarily the effects of the compounds have been taken into account however, where possible, practical criteria posed by the working procedures or the patterns of exposure which they determine have also been considered. Scientifically based criteria for health protection, rather than their technical or economical feasibility, are employed" [301].

It becomes evident from this statement that MAK-Values are primarily defined as 8-hour-averages. As the actual concentration of substances in the atmosphere of the workplace can vary within broad limits, 'Kurzzeitwerte' (Short-Time-Values) have additionally been introduced in order to assess the level and duration of excursions over the MAK. The respective substances have subsequently been grouped into five categories:

1. Substances which cause a local irritation effect;
2. Substances effective by resorption within 2 hours;
3. Substances effective by resorption after 2 hours;
4. Substances with a very low effect potential;
5. Substances with an intense odor.

Isocyanates, according to their mechanism of action, are included in category 1 for which the level of the 'Kurzzeitwert' is set at twice the MAK; the duration of each excursion must not be longer than 5 minutes; a maximum of eight excursions per working shift is admissible. It is also ruled that the twice

MAK level must under no circumstances be exceeded during the five minute period.

6.1.2 United States of America

The 'American Conference of Governmental Industrial Hygienists' (ACGIH) defines three different 'Threshold Limit Values' – TLVs [302]:

1. The 'Threshold Limit Value/Time-Weighted-Average' (TLV-TWA) – "the time weighted average concentration for a normal 8-hour workday and a 40-hour workweek, to which nearly all workers may be repeatedly exposed, day after day, without adverse effect". TLV-TWA values have been established for several isocyanates.
2. The 'Threshold Limit Value/Short Term Exposure Limit' (TLV-STEL) – "the concentration to which workers can be exposed continuously for a short period of time without suffering from 1) irritation, 2) chronic or irreversible tissue damage, or 3)......provided that the daily TLV-TWA is not exceeded......a STEL is defined as a 15-minute time-weighted average exposure which should not be exceeded during a workday even if the 8-hour time-weighted average is within the TLV. Exposures at the STEL should not be longer than 15 minutes and should not be repeated more than four times per day". There should be intervals of at least 60 minutes between successive exposures at the STEL.
   A TLV-STEL has only been stipulated to date by the ACGIH for 2,4-TDI.
3. The 'Threshold Limit Value-Ceiling' (TLV-C) is "the concentration that should not be exceeded during any part of the working exposure".
   The TLV-C can also be assessed by taking air samples over a 15-minute period if instantaneous sampling is not feasible excepting those substances which may cause irritation after exceedingly short exposures times.

The US 'Occupational Safety and Health Administration' (OSHA) not yet having implemented the ACGIH recommendations has itself issued 'Permissible Exposure Limits' (PEL) for 2,4-TDI and MDI.

6.1.3 France

The 'Institut National de Recherches Scientifiques' defines parallel to the TLV-TWA 'La valeur limite moyenne d'exposition' (VME) and 'La valeur limite d'exposition' (VLE); the latter is comparable to the TLV-STEL. The VME represents the mean of the concentration over an 8-hours working period; the VLE is taken as the average concentration over 5 minutes. A VLE equal to twice the VME is stipulated for several diisocyanates [303].

6.1.4 United Kingdom

The 'Health and Safety Executive' (HSE) in the UK has set 'Occupational Exposure Limits' and defines:

– 'Control Limits' "which have been judged after detailed consideration of the available scientific and medical evidence to be reasonably practicable for the whole spectrum of work activities in Great Britain. These.......limits ...should not normally be exceeded.......Failure to comply with control limits....may result in enforcement action";
– 'Recommended Limits' which "are considered to represent good practice and realistic criteria for the control of exposure.........HSE inspectors will use these...limits......for assessing compliance with the Health and Safety at Work Act.....".

All isocyanates – *NCO-bearing species* – fall under the category 'Control Limits'; in order to include products which have no definitive molecular weight, values are given in mg NCO/m^3 [304].

6.2 Exposure Limits for Diisocyanates

It is general practice to express the exposure limits of volatile substances in terms of ppm i.e. number of parts of vapor or gas per one million parts of contaminated air by volume at ambient temperature (20 . . . 25 °C) and a pressure of 1013 mbar (the equivalent concentration figures in mg/m³ are mostly communicated in parallel). In principle, this approach is only applicable to diisocyanates with a definite molecular weight. The definition of 'Control Limits' in the United Kingdom covers all NCO-bearing species regardless of whether they are airborne as a gas, vapor or aerosol and also includes mixtures containing polyols in a more or less progressed state of reaction. The obvious question of whether an exposure limit should generally be set for polyisocyanates also seems to be irrelevant if exposure limits for diisocyanates are observed since all NCO-groups are recorded by analytical determination and presented as diisocyanate, provided proper techniques for sampling and analytical determination of the airborne substances are used [305], see also [306,307].

The medical/scientific basis for laying down exposure limits is in general similar in different countries. It must, however, be emphasized that their status can vary between recommendations and legally binding requirements. The values have (see Tables 27a and 27b) been stipulated internationally at 0.005 . . . 0.02 ppm for the TWA (time- weighted average over the working shift) and usually at 0.02 ppm for the STEL (short term exposure limits as averages for 5 . . . 30 minutes). In several countries ceiling limits (C) have been set; a clear definition is often missing. The data represent the situation as it stood mid 1986 [308].

Table 27a. Workplace exposure limits for diisocyanates

Country	Substances	Exposure Limit (ppm)		Issued
Australia	TDI	C	0.02	1977
Austria	TDI, MDI, HDI, IDPI	C	0.02	
Belgium	2,4-TDI	TWA	0.005	1984
		STEL	0.02	
	4,4-MDI	TWA	0.02	
Bulgaria	HDI	C	0.007	1971
	TDI, MDI	No limits		
Canada	TDI, MDI	C	0.02	1980
CSSR	TDI	TWA	0.01	
		STEL	0.02	
	MDI	TWA	0.07	
		STEL	0.014	
Denmark	2,4- and 2,6-TDI, MDI, HDI, IPDI	C	0.005	1985
Finland	TDI, MDI	STEL	0.02	1982
	HDI, IPDI	STEL	0.01	1982
France	TDI, MDI, HDI, IPDI	TWA	0.02	
		STEL	0.01	1986
Germany FRG	2,4- and 2,6-TDI,	TWA	0.01	
	MDI, HDI, NDI, IPDI	STEL	0.02	1984
Germany GDR	TDI, HDI	STEL	0.007	
	MDI	TWA	0.015	
	MDI	STEL	0.015	
	IPDI	TWA	0.01	1978
	IPDI	STEL	0.01	1982
Hungary	TDI	TWA	0.014	
		STEL	0.028	
	HDI	TWA	0.007	
		STEL	0.014	1979
Italy	TDI, MDI	TWA	0.02	
	IPDI	TWA	0.01	
Japan	TDI	C	0.02	1965
	MDI	No limits		

6.3 Protective Measures in the Workplace

The main requirement for the maximum possible protection of employees is to keep the isocyanate concentration within the workplace below the exposure limits. It is rather easy to meet this requirement in production plants. As the industrial manufacture of these products is performed in *closed systems* escape of gaseous or liquid materials into the workplace atmosphere may occur only in accidental situations; personal respiratory protection devices should therefore be available. Respirators supplying pure air are preferred to those which only

Table 27b. Workplace exposure limits for isocyanates

Country	Substances	Exposure Limit (ppm)*		Issued
Netherlands	TDI, MDI	C	0.02	
	IPDI	TWA	0.01	
	HMDI	C	0.01	
Norway	TDI, MDI	TWA	0.005	
	HDI, IPDI	STEL	0.01	1984
Romania	TDI	TWA	0.014	
		STEL	0.043	
	MDI	STEL	0.015	
S. Africa	TDI, MDI	Usually follow ACGIH		
Spain	TDI, MDI	Usually follow ACGIH		
Sweden	TDI, MDI, HDI, NDI,	TWA	0.005	
	IPDI, TMDI	STEL	0.01	1984
Switzerland	2,4- and 2,6-TDI	TWA	0.005	
	MDI, HDI, IPDI	TWA	0.01	
U.S.A.	ACGIH recommendations			
	2,4-TDI	TWA	0.005	
		STEL	0.02	
	MDI	TWA	0.005	
	HDI	TWA	0.005	
	IPDI	TWA	0.005	
	HMDI	TWA	0.005	1988
	OSHA Permissible Exposure Limit (PEL)			
	2,4-TDI	C	0.02	
	MDI	C	0.02	
USSR	TDI, HDI	C	0.007	1976
Yugoslavia	TDI, MDI	C	0.02	1972
	HDI	C	0.007	
United Kingdom	–NCO containing products	TWA	0.02 mg/m^3	
		STEL	0.07 mg/m^3	
		(–NCO)		1983

* United Kingdom figures quoted in units of mg/m^3.

purify the air. Compliance of the respiratory equipment with the requirements of official bodies such as the Occupational Safety and Health Administration (OSHA) has to be observed [309].

The multitude of polyurethane manufacturing methods, in contrast, is associated with many different working environments in which the isocyanate is processed mostly in a more or less *open system*. Stationary continuously and discontinuously operating plants give rise to different exposure scenarios [310]; the situation can be kept rather easily under control by installing suitable exhaust facilities. In the case of non-stationary processing of isocyanates e.g. spray- and in-situ-foaming, conventional and spray application of coatings, personal protection for the worker must be provided; possibly, the simultaneous presence of solvents and other hazardous substances has to be taken into account. This is particularly relevant for tertiary aliphatic amines which are used as polyurethane reaction catalysts [311, 312]. For detailed information on precautionary measures, see [313]. Recommendations for improving industrial hygiene can be found in [128].

6.4 Monitoring in the Workplace

Figures on the concentration of hazardous chemicals in the workplace are an important basis for the medical assessment of the hazard presented by chemicals and the resulting recommendations and regulations implemented. Regular monitoring of exposure levels is essential for the protection of employees. The data have to be correct and representative:

– Figures which are too low because of incorrect monitoring endanger the health of the workforce:
– Erroneously high figures may lead to unnecessary and expensive technical prevention measures.

In either case false conclusions may result as to the medically acceptable exposure limit.

The first step in evaluating the exposure situation in a newly established workplace is the basic analysis of the working area. It may be necessary to employ specific and sophisticated measuring techniques to develop a reliable picture of the situation. The procedure

– has to be sufficiently specific in order to avoid falsification of the results because of the detection of other materials present;
– must ensure correct determination of the chemical in question – which is essential in analysing isocyanates because of their extremely low exposure limits;
– should guarantee a sufficiently reliable and satisfactory detection limit;
– should allow fool-proof execution.

After the basic evaluation, regular monitoring of the situation in the workplace is needed for which simpler procedures may be sufficient. For a proposal for a surveillance strategy in polyurethane foam plants, see [314].

In general, two methods exist for the determination of airborne isocyanates:

– the '*Wet*' – (*Two-Step*) – *Method*: the contaminated air is absorbed in a solvent containing a derivatizing reagent or adsorbed on the surface of a solid carrier material impregnated with such a reagent; the isocyanate forms a new compound which is determined in quantity by photometric or chromatographic methods;

– the '*Dry*' – (*Direct-Read-Out*) – *Method*: the isocyanate in the air sample is derivatized to a colored compound on a paper tape impregnated with a reagent; the color intensity indicates the concentration of the contaminant. Direct read-out of the concentration is also possible by a technique using a piezoelectric quartz crystal and by a method based on the ionization of the contaminant by a radioactive source.

Most of these procedures have primarily been developed for diisocyanates in vapor form (mainly TDI).

Polyfunctional isocyanates can as outlined above be present in the workplace in various forms. TDI and HDI will predominantly exist as a vapor, those with a lower vapor pressure e.g. MDI and polyisocyanates as a 'reacting' aerosol. Sampling techniques and analytical methods have to be adapted to suit these different physical states.

Several reviews summarize the actual state of the art [315–323].

6.4.1 'Wet' (Two-Step) Methods

The execution of this method requires – in contrast to 'direct-read-out' monitoring – a significant investment in analytical equipment and well trained and experienced personnel. The time which elapses between sampling and having the concentration figures at hand can vary between hours and days depending on the availability of equipment.

6.4.1.1 Sampling

The stream of contaminated air is generated by pumps with a capacity of approximately 2 l per minute. As the concentration of the contaminant to be determined is in the order of several $\mu l/m^3$, the minimum sampling time has to be 15 . . . 20 minutes for the majority of analytical methods, regardless of whether continuous or cumulative sampling is performed.

If the isocyanate exists in the air as a vapor, wash bottles containing the reagent solution which may be equipped with impingers can be used for sampling. Unbreakable absorption bottles and portable pumps are advisable if the equipment has to be worn by the employee in order to monitor the exposure level in his respiration area. Alternatively sampling tubes containing glass fiber filters and silanized glass wool respectively, which are impregnated with the reagent can be used. The isocyanate undergoes 'chemosorption' onto the glass surface; the derivative is dissolved in a suitable liquid and determined analytically [318, 324–328].

Several researchers have shown that quantitative absorption and derivatization in a wash bottle is not possible if the isocyanate is present as an aerosol [318, 325, 328, 329]. It is therefore better to carry out 'tube sampling' if the presence of NCO-bearing aerosols in the workplace air is anticipated. Isocyanate vapor will be derivatized simultaneously. It is advisable to combine both techniques by connecting a sampling tube to the outlet of a wash bottle.

6.4.1.2 Analytical Evaluation

Numerous methods have been proposed for the quantitative determination of the derivates obtained during sampling and for presenting the results in terms of isocyanate concentration in ppm or mg/m^3. They can be classed as *photometric* or *chromatographic* procedures.

Photometric Methods

The same basic principle is common for all photometric (colorimetric) methods: the diisocyanate is hydrolysed to the respective amine with acid: the amine is diazotized and coupled with a suitable reagent – mostly *N*-(1-naphthyl)-ethylenediamine – thus forming an azo-dye which can be determined photometrically [330–339]. It is obvious that this method is primarily only applicable for aromatic isocyanates; interference with aromatic amines which are sometimes used as auxiliaries in the manufacture cannot be excluded. A method for the determination of aromatic isocyanates in the presence of amines [340] is debatable. More recently a triazene compound with the chromophoric azo-group in the molecule has been proposed for the determination of TDI [341] which, in contrast to the aforementioned methods, is independent of variations in the 2,4/2,6-isomer ratio, see [288]. Chromophoric nature of the derivatizing agent is the prerequisite for the photometric determination of aliphatic diisocyanates after acidic hydrolysis; 1-fluoro-2,4-dinitrobenzene has been found to be a suitable material [342–345].

Chromatographic Methods

Various chromatographic methods have been proposed for the determination of airborne diisocyanates. High Performance Liquid Chromatography has gained the greatest importance.

Gas Chromatography (GC)

This method has mainly been recommended for underivatized isocyanates (TDI, HDI) after trapping in a cold tube or in a wash bottle containing a solvent [346, 347, 348]; direct air injection has also been described [349]. Some methods include a derivatization step [350–355]. See also [356].

Thin Layer Chromatography (TLC)

An important step forward from conventional photometric methods was the development of a secondary amine {*N*-(4-nitrobenzyl)-*N*-*n*-propylamine} as the derivatizing reagent. With diisocyanates it forms a stable urea derivative containing an aromatic nitro group; in further steps, excess reagent is removed after acidification in an aqueous environment, the nitro group is diazotized and coupled forming a brightly coloured dye [357]. Quantification was achieved by comparison with standard spots. This 'nitro-reagent-method' is applicable to aromatic and aliphatic isocyanates and mixtures thereof, adducts, prepolymers and other NCO-bearing species. A modification of the method related to sampling in a tube filled with glass powder wetted with the 'nitro-reagent' has been reported by the same authors [324].

1-(2-Pyridyl)-piperazine ('2-PP-reagent') is another secondary amine which has gained considerable importance as a reagent for forming a urea derivative with the isocyanate. It has primarily been used in TLC-determinations [358].

High Performance Liquid Chromatography (HPLC)
The 'nitro-reagent' method combined with HPLC can be used for most organic diisocyanates. Analysis of the urea derivative is performed by ultraviolet detection of the aromatic ring structures [359]. Difficulties with regard to the elution of excess reagent from the column were solved by a modification of the method [360]; efforts were made to lower the detection limits [361, 362] and to simplify the procedure [363]. The 'nitro-reagent'-method proved successful when used for the determination of isocyanates which were degradation products of polyurethane-coated wire [364]. The determination of IPDI was followed using the 'nitro-reagent', 1-(o-methoxymethyl)-piperazine and dibenzylamine as derivatizing agents [365].

The '2-PP-reagent' showed excellent sensitivity and performance when used as the derivatizing agent for subsequent HPLC-evaluation [366] and was particularly successfully used in TDI analysis [367, 368, 369]. Detection is achieved using ultraviolet radiation as well.

Derivation of the isocyanate with ethylalcohol is another approach used to analyse diisocyanates with HPLC and UV-detection [366, 370]. Because of the slow reaction of the alcohol, the method has its limitations. Diethylamine and aniline have also been studied as reagents [371, 372].

In order to increase sensitivity, several reagents were studied which seemed particularly suitable for detection by fluorescence. 1-Naphthylene methylamine [373], N-methyl-1-naphthylenemethylamine [374] and 9-(N-methylaminomethyl)-anthracene [375] were proposed for this technique. The latter reagent was used successfully for HDI tube sampling by chemosorption [376] and for TDI/MDI sampling in impinger equipped wash bottles containing the chemical as a solution in toluene [377]. Although the higher sensitivity allows a reduced sampling time, the method has found only limited application because of interference from other air contaminants.

For HPLC analysis involving electrochemical detection 1-(2-methoxyphenyl) piperazine [378] and p-aminophenol [379] have been proposed as derivatizing agents.

Mass Fragmentography
This method has been proposed for determining toluenediamine (TDA) formed by hydrolysis of TDI after sampling in an acidic medium [380].

6.4.2 'Dry' (Direct-Read-Out) Methods

Paper tape monitoring, although not specifically sensitive, is the classical way of achieving permanent and rather inexpensive workplace monitoring and control. Portable instruments of high sensitivity allowing the monitoring of TDI by ion mass spectroscopy may gain more significance in the future.

6.4.2.1 Paper Tape Monitors

The basic principle of paper tape monitoring for diisocyanates was developed more than 20 years ago [381]; appropriate devices were developed which assure a constant air stream taken into an opening where it comes into contact with the impregnated tape which moves continuously or discontinuously. The color forming reaction requires some time until completion.

The instruments available to date are easy to operate; they allow a follow-up of the concentration, if deemed appropriate, over a whole working shift but their use is limited to diisocyanates in the vapor state. Aerosols or particles because of uncontrollable adsorption in the sampling probe and irregular formation of spots on the tape often supply very unreliable data.

Commercially available monitors such as the MDA Model 7000 (stationary) and MDA Model 4000 MCM (portable) are suitable for aromatic isocyanates and in particular TDI; renewed calibration is required for other diisocyanates. The reaction system of the MDA Model 7005 also allows the determination of aliphatic diisocyanates. Whereas the tape of the aforementioned instruments works continuously, the MDA Model 7100 is equipped with a tape allowing stop-and-go operation. The supplier of the MDA-series of monitors can provide detailed information [382].

The portable SKC Monitor Model 920-01 Autostep is offered for monitoring TDI and MDI concentrations [383]. Critical evaluations of the techniques available and the result of comparative tests have been published by several authors [384–390]. It has been suggested that the readings of the MDA Monitors 7000 and 7005 depend to some extent on the ambient humidity which should consequently be recorded in parallel. The supposed formation of toluenediamine (TDA) from the hydrolysis of TDI and its interference with TDI readings on commercial paper tape did not receive closer study because researchers based their assumptions on false TDA data, see [297, 298, 299]. Changes of the TDI isomer ratio influence readings from the MDA 7005 tape; the MDA 7000 and SKC 920 tapes are not influenced adversely, see [288].

6.4.2.2 Piezoelectric Quartz Crystals

Piezoelectric quartz crystals coated with polyethylene glycol are suggested for use in determining TDI via surface adsorption and subsequent monitoring of the resulting weight increase by the associated change in the crystal oscillation frequency [391–394].

6.4.2.3 Ion Mass Spectroscopy

Air contaminants can be determined using mass-spectroscopy; the different velocities of the various ions in the spectrometer formed by irradiation is the basis used for their characterization. A model which has been specifically modified for TDI is being tested [395].

6.5 Discussion of Monitoring Methods

Discussion of the suitability of monitoring methods, their sensitivity and detection limits has to be made in close conjunction with the requirements on exposure limits which have been set on the basis of occupational health considerations. The magnitude of the figures presented in Table 27 is, in general, rather uniform but the individual definitions of what these figures really mean are rather confusing: "TWA"-values are usually defined as daily averages. "STEL"-figures are mostly averages over 15 minutes; however, 5 minute means (e.g. France, W. Germany, Sweden) and 10 minute means (e.g. U.K.) were also established. The definition of "C" (ceiling)-limits is very inaccurate and obviously often arbitrary.

Medical experience in contrast shows clearly that a maximum exposure level should never be exceeded – not even momentarily. 0.02 ppm is considered to be this limit which assures, under the given circumstances, the maximum possible protection of personnel.

It should not be especially difficult to approach this requirement in continuously operating diisocyanate production and processing plants. After a thorough basic analytical evaluation of the exposure levels, regular monitoring and control by a tape monitor (MDA 7000 for aromatic, also MDA 7005 for aliphatic types) would be sufficient, supposing the diisocyanate is present as the vapor. These stationary monitors run continuously; the time needed for the complete development of the color on the tape and the subsequent indication of the concentration varies: 15 minutes for the MDA 7000 and 1 minute for the MDA 7005. Accidental changes of the diisocyanate level can therefore be detected with the MDA 7005 after a rather short time. Alternatively, the stationary device MDA 7100 (color development requires 1 minute) can be used; it operates on a go-and-stop basis and can therefore be used for random sampling.

The portable types, MDA 4000 MCM (15 minutes development) and SKC 920 (1 minute), can be used for personal monitoring to be carried by employees who work at different locations in the plant and may be exposed to varying diisocyanate concentrations.

In plants with discontinuous operation and with concentrations varying between "highs" and "lows" e.g. production of moulded polyurethane articles, it should be confirmed by a basic study that the "highs", especially if they occur only instantaneously, do not exceed the 0.02 ppm level. This is done best by taking cumulative air samples in a wash bottle or sample tube at critical locations and during the periods at which the "highs" are most likely to occur. As the concentration of the diisocyanate may be in the order of 0.02 ppm a total sampling time of 10 . . . 20 minutes (2 l/min) is required if the evaluation is made photometrically. The sampling time can be reduced if chromatographic methods are employed but it has to be remembered that the particularly high sensitivity of HPLC with electrochemical or fluorescence detection may be disadvantageous because of interference from other air contaminants.

If this cumulative analysis of the "highs" results in values in the order of 0.02 ppm, tape monitoring, as described for continuous processing, can be used for control measurement. The tape may then show color shades of constantly changing intensities.

Tape monitoring is not possible for airborne diisocyanates if they are present as aerosols (possibly containing vapor) or if the air to be sampled contains aerosols or particles of other NCO-bearing species. Tube sampling, derivatization and photometric or chromatographic evaluation is in this case the only possibility for the basic evaluation and the regular control in the workplace. This problem may arise in pour-in-place and spray applications of foam and coatings out of plant. It is questionable whether, in this case, the sampling time can always be kept below the value which is prescribed to assess the compliance with the STEL (5–15 minutes) as cumulative sampling may not be appropriate due to the absence of permanently repeated operations.

These examples may suffice to illustrate the various methods for monitoring isocyanate concentrations in the workplace. It is clear that paper tape techniques are convenient and inexpensive but that they are limited to in-plant-use for diisocyanate vapor. The 'wet'-methods, as described in Section 6.4.1, are polyvalent but require laboratory facilities and – specifically for chromatographic evaluation – sophisticated and expensive equipment as well as qualified personnel.

The multitude of existing 'wet'-methods leads to another problem. If calibration is not made with sufficient care and if, for example, two methods have not been subject to comparative testing there is a great risk that they supply different values for identical exposure scenarios; such an outcome can lead to erroneous conclusions in the assessment of the health of workers. In view of the fact that research into more sensitive methods – mostly on the basis of complicated techniques – is continuing, it is recommended that

– a new method is carefully compared with several other procedures which can be considered as 'standard' and that it is only applied after publication in a scientific journal which includes the outcome of the comparative tests;
– an international industrial organization addresses the task to a group of experts to distill out a limited number of monitoring methods from the

published material particularly for areas which cannot be covered by paper-tape monitoring; these methods must be easy to handle by suitable plant personnel and should not require expensive and complicated equipment. International acceptance of these procedures should be mandatory for the industry.

6.6 Biological Monitoring

This subject although not covered by 'Monitoring at the Place of Work' seems to be worth mentioning in this section.

Two analytical studies are reported to identify TDI metabolites in the urine of rats which were exposed dermally to 2,4-TDI and in the urine of workers occupationally exposed to the 80:20 isomer mixture. The animal studies revealed that no TDI was present in the urine; 2,4-toluene diamine (TDA) was identified after acidic hydrolytic treatment [396]. It must be assumed that TDI is excreted as a conjugate which liberates TDA by hydrolysis [397], see also [398]. As the amount of TDA was in direct correlation to the estimated TDI dose the authors feel that this way of biological monitoring may be suitable for exposure control.

Conjugates which formed hexamethylene diamine (HDA) at acidic conditions were detected in the urine of workers handling HDI-based polyisocyanates in paint systems [399].

7 Occupational Health

This final section deals with the health hazards which could occur due to exposure to polyfunctional isocyanates and with the health injuries which could result therefrom. A great variety of observations, diagnoses and symptoms are the subject of publications in the medical literature; they are sometimes dazzling and confounding. For reasons outlined in [75] the author has confined himself to a general description of the symptomatology. A number of publications are quoted in this connection. The remaining references (from far more than 200 publications) resulting from a widespread literature search have been grouped under headings; they are listed in Section 7.8.

The number of employees handling polyfunctional isocyanates in their daily work has been the subject of several estimates in the U.S.A. According to a report presented to the Occupational Safety and Health Administration [400] about 230 000 employees were exposed to TDI and more than 100 000 to MDI in 1976. These figures are by far higher than the 1974 estimates of 50 000 . . . 100 000 workers being exposed to diisocyanates at any one time [401].

The discrepancies reflect the difficulties of realistically determining the number of people handling these chemicals in their workplace. It may however not be unrealistic to suppose that the number of persons working in the polyurethane industry worldwide is in the order of 500 000. This figure can obviously not include those who handle isocyanate based systems (e.g. coatings, adhesives etc.) for private do-it-yourself applications.

7.1 Exposure Situations

The exposure possibilities which may confront people when handling poly-functional isocyanates are:

Oral ingestion, although used routinely in animal tests as an administration route in determining the acute toxicity of a substance and often in assessing its chronic toxicity, would be unusual and occur only accidentaly for humans; this is particularly true of isocyanates since their strong irritating effect and pungent odor discourage swallowing;

Inhalation of vapor, in contrast, presents the most frequent exposure hazard. As pointed out in Section 4 the vapor pressure of diisocyanates vary greatly and accordingly so does the exposure situation. Polyisocyanates, provided that they contain only trace amounts of residual monomer, do not represent a specific isocyanate vapor exposure hazard; a possible organic solvent content, however, has to be taken into consideration.

Inhalation of particles is physiologically seen as a very complex process. Their diameter and size determines whether they can be considered as "respirable" [402]; their composition can vary within a broad range.

The following classification is applicable for the different types of particles:

– Liquid or nearly solid aerosols or droplets generated by spray techniques from a polyfunctional isocyanate/polyol reactive mixture as they are used in spray foam and coating applications; depending on the activity of the system it can be in a more or less progressed state of reaction; only the physiological effect of the NCO-group can be defined; although this is the most important type of aerosol occurring in practice, it is practically impossible to design an animal study to simulate the correct exposure conditions;
– Liquid or solid aerosols or droplets formed by condensation of diisocyanate vapor on cooling; this only rarely happens;
– Solid dust particles carrying mostly unreacted isocyanate on their surface; such particles can be formed in particular in the particle board bonding process with MDI, see 4.2.3.

Exposure by *contact with skin and mucous membranes* may occur because of insufficient personnel protection or accidental spillages; irritative and corrosion effects as well as toxicity following absorption have to be considered.

7.2 Exposure Levels in the Workplace

TDI levels measured in a manufacturing plant in the U.S.A. by the Marcali method [330] between 1986 and 1974 showed a progressive reduction from approximately 0.06 to less than 0.004 ppm [403]. In another plant, 8-hour TWA-levels of 0.0001 to 0.025 ppm, obtained by continuous tape monitoring, were reported [404]. A survey of the environmental situation in a TDI plant is given in [294]. The TDI concentration (determined according to Marcali) in air samples collected in a polyurethane plant dropped from an average TDI concentration of ≤ 0.03 ppm to < 0.008 ppm after the improvement of the exhaust system [405, 406]. A downward trend in TDI levels from 0.003 to 0.001 ppm during 10 years was also observed in two other polyurethane plants [407, 408, 409]. Peak concentrations were not measured in these studies. A 3-year survey of TDI concentrations in two foam plants resulted in TWA's of 0.0024 ppm, 0.0011 ppm and 0.0015 ppm for foam workers, finishing and maintenance personnel [410]; the use of a sequential tape monitor revealed levels in excess of the OSHA ceiling in 1.3% of the 12-minute samples [390]. Exposure limits were often exceeded temporarily by TDI and MDI in connection with moulding perations [411]. It is quite clear that spray application of polyurethane systems may lead to the surpassing of these limits.

Little information is available on the exposure situation in manufacturing plants for other isocyanates. HDI levels which occurred during the change of filters (except one) were ≤ 0.01 ppm [412]; MDI concentrations were hardly detectable with the exception of areas where dust from solid MDI flakes was generated [413]. NDI levels of $0.08 \ldots 2.2$ mg/m^3 determined by personal air sampling (particles, MAK: 0.09 mg/m^3) were responsible for initiation of an immediate improvement in occupational hygiene measures [414].

It is generally experienced that the exposure situation can be rather easily controlled in stationary production situations. Much more effort is needed to achieve compliance with occupational hygiene requirements for "on the spot" applications of isocyanates or polyurethane systems. Individual protection of the employee is often the most appropriate measure.

7.3 Exposure Scenarios and Arguments on the Health Status of Workers

When reading publications on occupational diseases which are attributed to isocyanate exposure, one has to note that the complexity of parameters which contribute to the overall exposure situation is often disregarded or not even taken into account. Subsequently incorrect interpretations of the symptomatology occurs or false conclusions on the real contribution of the isocyanate to the diagnosed illness are drawn. This is illustrated in the following:

– The authors of [415] regret that although numerous studies on the suggested effect of isocyanates on the lung function have been published, none of them present sufficiently precise data on the exposure situation. Due to the inadequate quantification of isocyanate exposure, a number of studies are not considered suitable to form an opinion of the lung function decrease caused by isocyanates. These studies were related to foam plant [416–425] and car plant workers [426,427].

– The findings of a study on a group of females employed in sewing seat covers containing polyurethane foam in a car plant are of little value as there are serious doubts as to whether the subjects were really exposed to TDI [428].

– Neurological complications like euphoria, ataxia and loss of consciousness on a group of firemen who were severely exposed to the combustion gases from a fire in a foam plant are repeatedly quoted in the literature and related to inhalation of TDI because a TDI tank leaked during the fire [429,430]. This statement is misleading as comparable symptoms have never been observed in other cases of massive TDI exposure; it disregards the fact that the *mixture* of volatile gases which are generated in a polyurethane fire (or possibly lack of oxygen) is responsible for the observed symptoms and not the TDI exposure as such.

– Toluene diamine (TDA) is, according to [129], specified as a "breakdown product" of TDI; it is stated that this chemical in contact with water "may be converted to the corresponding diaminotoluenes". This sweeping statement cannot be sustained for every type of contact of TDI with an aqueous environment and even less so for the gaseous phase (see Section 5.2).

– As respiratory tract irritation had been observed after inhalation of tertiary aliphatic amine catalysts, it was felt that TDA, as the supposed hydrolysis product of TDI, could cause such symptoms on foam workers as well; it was not noted that TDA is a primary aromatic amine with no irritation potential when inhaled. A comparative inhalation challenge test with the two substances confirmed the absence of an irritation effect by TDA [431, 432].

It becomes evident from these cases that the following prerequisites relating to the exposure situation are essential when injuries to the health of humans who are handling isocyanates are to be diagnosed:

– the researcher should have a sufficient basic knowledge of the chemistry of isocyanates in general and particularly of those which are prevalent in the workplace; there should be information available on the physical status of the airborne isocyanate; it should be ascertained whether, and if any which, other air contaminants are present; above all the presence of respiratory irritants like tertiary aliphatic amines and of solvents should be looked for;

– methods for air sampling and for determining the concentration levels of isocyanates must be reliable and have to comply with their chemical identity and physical status; this evaluation must be based on an appropriate "measuring strategy" which takes the exposure levels at the various locations of the plant into account; continuous sampling has to be complemented by cumulative sampling if fluctuations of the concentration levels are obvious; only if

figures on time-weighted-averages are accompanied by values on peak-con-
centrations (if any) a realistic assessment of the exposure scenario can be made;
it is self-evident that the concentration of other critical substances is also of
interest;
- it is of utmost importance to establish a "historical" picture of the development of exposure levels in the respective plant.

The assistance of an experienced industrial hygienist when planning and
executing these evaluations is helpful.

7.4. Oral Ingestion

According to a report on a suicide attempt, a 21-year old woman who was in the
24th week of pregnancy drank 60 ml of a hardener containing 10 ml of
a triisocyanate. She gave birth to a healthy boy 4 months later [433]. The
swallowing of a reacting polyol-MDI-mixture has also been reported [434].

7.5 Effects on Conducting Airways

Diisocyanates can cause toxic-irritative and sensitization effects. A spectrum of
various clinical pictures on the respiratory tract have resulted therefrom. The
following classification is based on a recommendation proposed in [435]. The
literature quoted in Section 7.5 includes the pertinent references specified in
[415] and [435]. They relate mainly to cases of TDI exposure.

7.5.1 Toxic Irritation

Depending on the seriousness of the exposure, acute, subchronic or chronic
clinical effects will develop in all involved subjects: 0.05 ppm (30 minutes
exposure) leads to lachrymation and running nose; 1.3 ppm (for 10 minutes)
causes in addition pharyngitis and bronchitis; from 50 ppm the situation
becomes very critical as the danger of lung edema arises. Repeated exposure to
concentrations above 1 ppm can lead to a subchronic irritation syndrome.
A chronic irritation syndrome can develop after exposure to levels exceeding
0.02 ppm: dyspnoe and a reduction of the lung function can occur [436], see also
[437–445]. Cases of slight and transient irritation have been found on numerous
occasions. The number of incidents caused by massive inhalation of diisocyanates followed by acute poisoning is limited [446–449].

The "Immediately Dangerous to Life or Health (IDLH) Concentration" i.e.
the maximum level from which one could escape within 30 minutes without any

escape-impairing symptoms or any irreversible health effects has been set for TDI at 10 ppm [450]. This has to be considered with respect to the strongly pungent odor of this chemical: It can be noticed at a concentration of 0.03 . . . 0.05 ppm and becomes intolerable from 3 . . . 5 ppm [177].

7.5.2 Effects on Pulmonary Function

Subchronic and chronic irritation as well as chronic bronchitis can naturally lead to disorders of the respiratory function. It is still a subject of debate whether additionally a lingering impairment of the lung function can occur which is not accompanied by subjective discomfort. In the majority of epidemiological studies no deterioration of the average lung function has been found as long as a ceiling concentration of the diisocyanate is not – or only incidentally – exceeded [443, 451–462].

According to [457–463] lung functions following acute exposure i.e. during a working shift at low or slightly elevated concentrations, remain unchanged. There are some data reported in the literature which are contrary to these findings [405, 406, 464–466]. These publications describe a significant decrease of the lung function during a working shift at TDI levels of 0.014 ppm (average of 1 hour). Serious doubts against these investigations however have been voiced [404, 436, 467].

Cross section epidemiological studies can supply information on a possible chronic influence of isocyanates on the lung function. The following papers report normal values of the lung function although the 0.02 ppm level was temporarily exceeded in the workplace [413, 443, 456, 457, 460]. At higher concentrations reduction of the lung function compared to control groups or to literature data has been observed [411, 461, 462].

Longitudinal epidemiological studies also give evidence to possible chronic effects on the lung function. No significant decreases compared to the standard at low and moderately elevated concentrations are reported in [409, 413, 454, 457, 460, 468, 469]. Partially significant annual decreases of the lung function at concentrations below 0.02 ppm were only described in [464, 465, 470–472]. A group which was only occasionally exposed to concentrations above 0.02 ppm showed a deterioration of the lung function; the averages were obviously influenced by single persons with strongly damaged respiratory organs [453]. A significant annual decrease occurred at strongly elevated levels of over 0.5 ppm [462].

7.5.3 Specific Isocyanate Hypersensitivity

The formation of hypersensitivity is, as observations on humans and results from animal studies indicate, linked to the exposure threshold: it can be assumed that in the case of strictly adhering to a concentration of ≤ 0.02 ppm a development of hypersensitivity must not be expected on normal healthy persons [223, 403, 473, 474]. Although it is not fully clear whether a predisposition exists for humans, it is a matter of fact that at the time only some individuals out of a larger group develop a hypersensitivity in the workplace if overexposed to the same degree. Immunological or metabolic individualities may be the cause; possible skin contact has to be taken into account also, as pulmonary sensitization after dermal exposure has been confirmed in an animal study [234].

In the case that a specific hypersensitivity of the conducting airways already exists, very low isocyanate levels (0.01 . . . 0.02 ppm, in specific cases possibly 0.001 ppm) are sufficient to initiate distinct clinical circumstances showing two typical symptomatologies.

7.5.3.1 The "Isocyanate-Asthma"

The clinical picture of isocyanate-asthma is not uniform. In the initial stage shortness of breath, difficulty of breathing and tickle in the throat can occur in the case of reexposure. If exposure is continued typical asthmatic attacks (of immediate, dual and delayed type) can develop; alternatively a chronic-obstructive bronchitis can be formed. Pharmacological [57, 475–477] as well as immunological [234, 248, 478–483] mechanisms are discussed for this clinical picture; see also [239, 484–486] and [62, 252, 487–489].

Specific isocyanate-asthma must be delimitated by differential diagnosis from the subchronic and chronic irritation syndrome as well as from obstructive lung diseases of other origin. A detailed anamnesis and a specific evaluation of the exposure situation are of utmost importance. Control of the Vital Capacity and challenge with acetylcholine [490, 491, 492] can positively contribute to the diagnosis. Immunological examinations like the radio allergo sorbens test (RAST) yield evidence if they are positive; this has been stated only in 10 . . . 50% of the cases [478, 493]. A provocative inhalation challenge test (PIC) with the isocyanate in question would be the most important criterion; the carrying out of this test is naturally limited to locations possessing the necessary equipment and qualified personnel.

A recovery from isocyanate asthma can be expected if the exposure is stopped at an early stage. Only severe cases can lead to chronic bronchitis [417, 494].

7.5.3.2 The Allergic Alveolitis

The allergic alveolitis or "hypersensitivity pneumonitis" is another manifestation of a specific isocyanate hypersensitivity which has been observed only in a dozen cases (about 50% were diagnosed on MDI exposed individuals). The clinical picture is comparable with the farmers- or threshers disease: sensitization develops after exposure to the allergen over many years. Reexposure can afterwards initiate an influenza type of illness followed by fatigue, rheumatism, fever, shivers, dyspnoe, possibly also nausea and vomiting. Combinations of these symptoms with asthmatic reactions and the transition to a fibrosis have been described in only a few cases [495–502].

7.5.3.3 Observations on Hypersensitive Individuals

It has been reported that isocyanate asthma originates predominantly from massive overexposure [451, 469, 490, 503, 504]. Several cases of asthma are described which were related to average TDI exposure levels above 0.03 . . . 0.06 ppm; in these studies no bronchial sensitization was developed at average concentrations below 0.02 . . . 0.03 ppm [403, 437, 473, 505]. Some reports describe sensitization effects on individuals at even lower concentrations [506, 507, 508]; the exposure levels in these cases were only estimated or only peak concentrations were not taken into account.

75% of individuals with bronchial irritation symptoms and a distinct decrease of the respiratory capacity after an isocyanate incident showed a positive RAST; those solely with irritation symptoms showed a positive RAST only in 10% of the cases [474]. Cohorts with typical exposures below 0.02 ppm but to momentary peaks up to 1.3 ppm showed throughout a negative RAST [248, 482, 509].

7.5.4 Unspecific Bronchial Hypersensitivity

It estimated that approximately 10 . . . 15% of the non isocyanate exposed population is sensitive to numerous physical and chemical noxas (pollen, wood dust, cold air etc). Isocyanate asthma is in the majority of cases accompanied by this unspecific hypersensitivity [475–477, 492, 510, 511]. Persons having trouble with such unspecific hypersensitivity will conversely react to a lower isocyanate concentration (≥ 0.01 ppm) with a bronchospasm. These individuals should therefore not handle isocyanates [512].

7.6 Effects of Dermal Contact

Several clinical human cases have been described [513]. One author reports twenty cases of occupational dermatoses caused by different isocyanates [514]. A case of dermatitis has been related to contact with an MDI- containing solution [515]. Dermal contact was felt to have initiated a number of cases of occupational asthma and respiratory sensitivity [516–519]. Contact dermatitis related to HDI is reported in [520].

Contact dermatitis was observed in workers in a plant in which HMDI based polyurethane was handled; patch testing revealed positive reactions suggesting sensitization to this substance [521]. Contact sensitivity to HMDI was reported in [522]. Various clinical syndroms following skin contact were observed in workers handling HMDI based coatings. The signs of intoxication disappeared after 10 to 14 days from the beginning of appropriate medicinal treatment [523].

7.7 Medical Supervision

Medical supervision of all subjects who handle or come into contact with polyfunctional isocyanates is strongly recommended. This supervision should include preemployment screening, regular examinations and health control after absence for sickness.

7.7.1 Pre-Employment Screening – Medical Examinations

A detailed procedure for a medical screening prior to employment has been proposed by the professional association of the industry in West Germany [524]. This should consist of an anamnesis with particular emphasis on the respiratory system and a clinical examination including respiratory function tests. Persons with asthmatic-type conditions, chronic bronchitis or other chronic respiratory diseases, recurrent eczema, skin allergies and hay fever should be excluded from working with isocyanates. There is at present no screening test suitable for detecting persons who may become sensitized.

As a rule a proportion of subjects who become sensitive to isocyanates will develop symptoms during the first six months. Respiratory function tests are therefore recommended 6 . . . 12 months after commencement of handling. Thereafter, medical examinations should routinely be made at intervals of 1 . . . 2 years. It is prudent to examine all cases of prolonged absence due to sickness related to the respiratory tract.

7.7.2 First Aid and Therapeutic Measures

The leaflets of the suppliers of polyfunctional isocyanates contain detailed recommendations on actions to be taken in cases of accidental overexposure. If the chemical has entered the eyes, flushing with luke-warm water for several minutes is essential. Contaminated skin should be washed with soap and water. Ingestion of the material should be treated by giving 250 ml milk or water to drink if the patient is conscious and transferring him to a medical facility for gastric lavage.

In case of inhalation the affected person should be removed as soon as possible to fresh air and kept at rest. Asthmatic type symptoms may develop; they could be delayed for up to 12 hours. The medical treatment is essentially symptomatic. Researchers in Italy have in this connection studied the suitability of several drugs [525–530].

7.8 Relevant Medical Literature

The literature search made for Section 7 of this monograph resulted in more than 200 references from which about 100 were included in the preceding paragraphs. The author feels that the remaining literature should be cited as the interested scientist may wish to study the international situation in this field in more detail. In order to simplify the presentation, the references were grouped under various generic headings to which one reference number relates:

- General papers and review articles [531];
- Epidemiological studies on individuals in TDI foam plants [532];
- Epidemiological studies on individuals in TDI plants [533];
- Epidemiological studies on individuals handling MDI [534];
- Epidemiological studies on individuals handling components for polyurethane coatings [535];
- Miscellaneous aspects of epidemiology [536];
- Provocative inhalation challenge – techniques and experiences [537];
- Mechanisms of development of isocyanate hypersensitivity and asthma [538];
- Case histories [539].

Acknowledgment The author is indebted to Akzo Chemie BV, Bayer AG, Hüls AG, Takeda Chemical Industries, Ltd. and The International Isocyanate Institute Inc. for submitting information and giving valuable advice.

8 References

1. *Wurtz A* (1849) Ann 71:326
2. *Pinner S H* (1947) Plastics 11:257
3. *Bayer O, Rinke H, Siefken W, Orthner L, Schild* (1942) Ger Pat 728981
4. *Bayer O* (1947) Angew Chem 59:275
5. SRI International (1987) Chemical Economics Handbook, Diisocyanates and Polyisocyanates, p 666–5021
6. SRI International (1987) Chemical Economics Handbook, monoisocyanates, p 666–6050
7. International Isocyanate Institute Inc., 119 Cherry Hill Road, Parsipanny NJ 07054 USA
8. *Saunders J H, Slocombe R J* (1948) Chem Rev 43:202
9. *Siefken W* (1949) Ann 565:75
10. *Arnold R G, Nelson J A, Verbanc J J* (1957) Chem Rev 57:47
11. *Saunders J H, Frisch K C* (1962) Polyurethanes Part I, Interscience Publishers, New York, p 180
12. *Eisenmann K H, Zenner K F* (1977) In: Ullmann, 4th edn, vol 13 p 347
13. *Findeisen K, König K, Sundermann R, Ippen J, Goerdeler J* (1983) Isocyanates. In: Houben-Weyl, Thieme, Stuttgart, vol E4 p 738
14. *Schauerte K* (1985) Polyurethanes. In: G Oertel (eds) Plastics Handbook. Carl Hanser, Munich, vol 7 p 42
15. *Dieterich D* (1987) Polyurethane, Polyharnstoffe Polycarbodiimide. In: Houben-Weyl (ed) Thieme, Stuttgart, vol E20 part 2 p 1561
16. *Hentschel W* (1884) Ber 17:1284

17. *Gattermann L, Schmidt G* (1888) Ann 244:30
18. *Ulrich H, Tilley J N, Sayigh A A R* (1964) J Org Chem 29:2401
19. *Ulrich H, Tucker B, Sayigh A A R* (1968) J Org Chem 33:2887
20. *Twitchett H J* (1974) Chem Soc Rev 3:209
21. *Morschel H, Skopalik C* (1963) Bayer AG: DAS 1 154 090
22. *David D J, Staley H B* (1969) Polyurethanes part III: Anal Chem of Polyurethanes, Wiley Interscience, New York
23. ASTM D 1638–74: Urethane Foam Isocyanate Raw Materials
24. ASTM D 1768–73: Toluene diisocyanate
25. *Arnold R G, Nelson J A, Verbanc JJ* (1957) Chem Rev 57:47
26. *Saunders J H* (1959) Rubber Chem Technol 32:337
27. *Saunders J H, Slocombe R J* (1948) Chem Rev 43:203
28. *Petersen S* (1949) Ann 562:205
29. *Brochhagen F K, Dieterich D* (1982) I.I.I. Project E-B 23
30. *Fraenkel-Conrat H, Cooper M, Olcott H S* (1945) J Am Chem Soc 67:314
31. *Schick A F, Singer S J* (1961) J Biol Chem 236:2477
32. *Fasold H, Turba F* (1963) Biochem Zeitschr 337:80
33. *Fasold H* (1965) Biochem Zeitschr 342:288
34. *Ozawa H* (1967) Biochem 62:419
35. *Sanderson C J* (1970) Immunol 18:353
36. *Petit L, Gervais P, Denuc AM* (1978) Bull Med Leg Toxicol 21:759
37. *Tse T S T, Pesce A J* (1979) Toxicol Appl Pharmacol 51:39
38. *Karol M H, Hauth B A, Alarie Y C E* (1979) Toxicol Appl Pharmacol 51:73
39. *Baur X* (1986) Allergologie 11:487
40. *Scheel L D, Killens R, Josephson A* (1964) Am J Ind Hyg 25:179
41. *Brown W E, Wold F* (1971) Science 174:608
42. *Twe J, Wold F* (1973) Biochem 12:381
43. *Brown W E, Wold F* (1973) Biochem 12:828
44. *Brown W E, Wold F* (1973) Biochem 12:835
45. *Twu J, Chin CCQ, Wold F* (1973) Biochem 12:2856
46. *Snyder P D, Wold F, Bernlohr R W, Dullim C, Desnick R J, Krivit W, Condie R M* (1974) Biochim et Biophys Acta 350:432
47. *Krupka R M* (1974) Pestic Sci 5:211
48. *Basil B B, Basil E F, Laszlo J, Wheeler G P* (1975) Can Res 35:1
49. *Brown W E* (1975) Biochem 14:5079
50. *Gross M, Whetzel N K, Folk J E* (1975) J Biol Chem 250:7693
51. *Ardelt W, Koj A, Chudzik J, Dubin A* (1976) FEBS Let 67:156
52. *Ogata F, Ninomiya N, Makisumi S* (1978) Faculty of Science Kyushu Univ Ser C 10:205
53. *McKenna R, Ahmad T, Frischer H* (1978) Clin Res 26:669
54. *Ahmad T, Frischer H* (1978) Clin Res 26:701
55. *McKenna R, Ahmad T, Frischer H* (1979) Thromb Haemost 42:12
56. *Trevisan A Q, Moro G* (1981) Int Arch Occ Envir Health 49:129
57. *Brown W E, Green A H, Karol M H, Alarie Y C E* (1982) Toxicol Appl Pharmacol 63:45
58. *Dewair M, Baur X, Fruhmann G* (1983) J Occ Med 25:279
59. *Brown W E, Green A H, Cedel T E, Cairns J* (1987) Env Health Persp 72:5
60. *Baur X, Dewair M, Fruhmann G* (1982) Eur J Resp Dis 62:113
61. *Stokinger H E, Mountain J T, Scheel L D* (1968) Ann N Y Acad Sci 151:968
62. *Avery S B, Stetson D M, Pan P M, Mathews K P* (1969) Clin Exp Immunol 4:585
63. *Taylor G* (1970) Proc Roy Soc Med 63:379
64. BEICIP Bureau d'études industrielles et de coopération de l'Institut Français de Pétrole Paris (1985) Aromatic Intermediates Report 87057-1985
65. Union Carbide Corp (1976) Methylisocyanate F 41443A 7/76
66. *Andersson N, Kerr Muir M, Salmon A G, Wells C J, Brown R B, Purnell C J, Mittal C J, Mehra V* (1985) Lancet 1:761
67. *Hofmann A, Neufelder M* (1971) Arch Toxicol 29:73
68. *Lohs K, Swart H, Junghans A* (1985) Zeitschr für Chemie 25:197
69. DIN Safety Data Sheets 014803/02, 0f9813/01, 014722/00, 013211/01, 019775/01 (24 March 1988), 019171/02 (26 August 1988), 014749 (8 December 1988) Bayer AG Business Group Organic Chemicals

70. DIN Safety Data Sheets 011960/02 (10 September 1987); 017241/01, 017950/01 (14 December 1987); 011367/01, 014854/02 (16 December 1987) Bayer AG Business Group Organic Chemicals

71. DIN Safety Data Sheet 017411/01 (24 March 1988), 013424 (26 August 1988) Bayer AG Business Group Organic Chemicals

72. Determined according to DIN 51755 (n-propyl-, i-propyl-, i-butyl-, t-butyl-IC); DIN 51758 (others)

73. Determined according to DIN 51794

74. *Salmon A G* (1985) Brit J Ind Med 42: 577

75. As the author has no medical training and only a general knowledge of toxicology, he confines himself to quoting rather than interpreting respective information from the scientific literature

76. *Kimmerle G, Eben A* (1964) Archiv für Toxikol 20: 235

77. *Vernot E H, MacEwen J D, Haun C C, Kinkead E R* (1977) Toxicol Appl Pharmacol 42: 417

78. *Procter N H, Hughes J P* (1978) Chemical Hazards of the Workplace, Lippincott, Philadelphia

79. *Lee C K* (1977) J Biol Chem 36: 754

80. *Pozzani U C, Kinkead E R* (1966) American Ind Hyg Conf, Pittsburgh PA

81. Methyl Isocyanate: Handling – Security – Storage (1970) Progil, Paris

82. Handbuch zum Umgang mit Methylisocyanat (1978) Bayer AG

83. American Conference of Governmental Industrial Hygienists ACGIH (1971) Documentation of the Threshold Limit Values, 3rd edn

84. Deutsche Forschungsgemeinschaft (1964) Mitteilung V der Senatskommission zur Prüfung gesundheitsschädlicher Arbeitsstoffe, Verlag Chemie, Weinheim

85. *Quinot E, Moncelon B, Millard M* (1978) Cah Notes Doc 93: 547

86. *Morel C, Gendre M, Limasset J C, Cavigneaux A, Protois J C* (1981) Cah Notes Doc 104: 451

87. *Neff J E, Ketcham N H* (1974) Am Ind Hyg Ass J 35: 468

88. *Malceva A S, Kogan L M, Frolov J E* (1968) Z Vses Chim Obsc 13: 230

89. *Back K C, Thomas A A, MacEwen J D* (1973) Aerospace Med Res Lab Wright, Patterson Air Force Base, Ohio PB-225283/1

90. Bhopal Methylisocyanate Incident Investigation Team Report (1986) Union Carbide, Danbury CT

91. *Singh M P, Ghosh S* (1987) J of Hazard Mat 17: 1

92. *Bucher J R* (1987) Fundam Appl Toxicol 9: 367

93. Envir Health Persp (1987) p 1–309

94. *McConnell E E, Bucher J R, Schwetz B A, Gupta B N, Shelby M D, Luster M I, Brody A R, Boorman G A, Richter C* (1987) Envir Sci Technol 21: 188

95. *Nemery B, Dinsdale D, Sparrow S* (1987) Bull Eur Physiopathol Resp 23: 315

96. *Diller W F* (1985) Dtsch Med Wochenschr 110: 1749

97. *Daunderer M* (1985) Münch Med Wschr 127: 24

98. Reported in EPA TSCA Chemical Inventory (1983)

99. *Gurova A I, Alekseeva N P, Gorlova O E, Chernyshova R A* (1976) Gig Tr Prof Zabol 3: 53

100. *Engelhard H* (1976) Zeitschr für die Gesamte Hygiene 22: 235

101. *Karol M H, Dixon C, Brady M, Alarie Y* (1980) Toxicol Appl Pharmacol 53: 260

102. *Sangha G K, Matijak M, Alarie Y* (1981) Toxicol Appl Pharmacol 57: 241

103. Reported in EPA TSCA Chemical Inventory (1986)

104. (1980) Agric Biol Chem 44: 80

105. National Cancer Institute (1986) Screening Programme Data Summary

106. *Yamaguchi T* (1980) Agric Biol Chem 44: 3017

107. *Renman L, Sangö C, Skarping G* (1986) Am Ind Hyg Ass J 47: 621

108. *Frolova I N* (1967) Gig Tr Prof Zabol 11: 23

109. *Zibireva I A* (1967) Gig Sanit 32: 3

110. *Frolova I N* (1966) Gig Sanit 31: 108

111. *Frolova I N* (1966) Gig Sanit 31: 481

112. *Voznesenskaya G A* (1968) Gig Trud Prof Zabol 12: 59

113. *Lomonova G V* (1969) Gig Tr Prof Zabol 13: 50

114. *Oshawa T, Ishizu S* (1983) Tokyo Joshi Ika Daigaku Zasshi 53: 273

115. Robra: tested according to Robra K H (1979) Vom Wasser, Verlag Chemie, Weinheim, vol 53 p 267

116. Bringmann: Tested according to Bringmann G, Kühn R (1977) Z für Wasserund Abwasserforschung 10: 87

117. Fish toxicity tested according to ISO DP 8912 = OECD 209
118. *Kaiser K L* (1985) Water Poll Res J Can 20: 38
119. *Janicke W* (1983) WaBoLu Ber 1/83
120. *Houk V S, DeMarini D M* (1988) Envir Molec Mutagen 11: 13
121. *Mukherjee A K* (1985) J Econ Taxon Bot (Allahabad India) 7: 568
122. *Saify T, Bhat P K* (1985) J Sci Res (Bhopal India) 7: 17
123. *Prasad R, Pandey R K* (1985) J Trop For (India) 1: 40
124. *Raza Khan M, Iqbal S A, Chaghtai S A, Saify T, Husain I* (1986) Indian J Appl Pure Biol 1: 47
125. The designatioh "Polyfunctional Isocyanates" includes diisocyanates and products of higher functionality, the term "Polyisocyanates" designates compounds with a functionality higher than two.
126. Bender D (1987) Keynote presentation at the Polyurethane World Congress, Aachen
127. I.I.I. Technical Information No 1 (1980) Recommendations for the handling of toluene diisocyanate (TDI)
128. *Woolrich P F* (1982) Am Ind Hyg Ass J 43: 89
129. World Health Organization IARC Monograph (1986) Toluene Diisocyanate, WHO, Geneva, vol 39 p 287
130. World Health Organization Environmental Health Criteria 75 (1987) Toluene Diisocyanates
131. Recommendations for the Handling of Desmodur[R] T Toluene Diisocyanate (1985) Bayer AG Business Group Polyurethanes
132. Environmental Protection Agency (1984) Chemical Hazard Information Profile Toluene Diisocyanate – TDI (Draft)
133. *McFadyen P* (1976) J Chromatogr 123: 468
134. *Bagon D A, Hardy H L* (1978) J Chromatogr 152: 560
135. *Taymaz K* (1986) L Liq Chrom 9: 3347
136. *Hannemann W W, Robison L L* (1968) Gas Chromatogr 6: 256
137. Environmental Protection Agency (1984) Chemical Hazard Information Profile, MDI
138. I.I.I. Technical Information No 4 (1982) Recommendations for the Handling of 4,4'-Diphenyl-methane Diisocyanate Monomeric and Polymeric
139. *Gudehn A* (1985) Am Ind Hyg Assoc J 46: 142
140. *Brochhagen F K, Schal H P* (1986) Am Ind Hyg Ass J 47: 225
141. DIN Safety Data Sheet Desmodur[R] 15 04805/01 (August 1988) Bayer AG Business Group Polyurethanes
142. NEL 03-8511 Elate[R] Diisocyanate (1985) and Product Safety Data Leaflet Elate[R] PPDI (1984, updated 1988), Akzo Chemie
143. NEL 04-8807 Elate[R] CHDI (1988) and Product Safety Data Leaflet Elate[R] CHDI (1984, updated 1988), Akzo Chemie
144. DIN Safety Data Sheet 303686/02 Desmodur[R] H (March 1988) Bayer AG Business Group LS
145. Produktinformation TMDI Trimethyl-hexamethylendiisocyanat (1986) and DIN Sicherheits-datenblatt TMDI (1985) Hüls Aktiengesellschaft
146. *Kanzawa T, Naito K* (1967) Japan Chemical Quarterly III–IV: 38
147. DIN Safety Data Sheet 028979/03 Desmodur[R] W (April 1988) Bayer AG Business Group LS
148. DIN Safety Data Sheet 303708/02 Isophorone Diisocyanate (IPDI) (April 1988) Bayer AG Business Group LS
149. Produktinformation IPDI Isophorondiisocyanat (1986) and DIN Sicherheitsdatenblatt IPDI (1985) Hüls Aktiengesellschaft
150. *Tanaka M, Nasu H* (1980) International Progress in Polyurethanes, Technomic Publishing, Westport CT USA, vol 2 p 27
151. *Kazama S* (1980) J Takeda Res Lab 39: 202
152. The designation "Polyurethane" has become customary for polymers based on diisocyanates regardless of urea, biuret, allophanate, carbodiimide, polyisocyanurate or other isocyanate related bonds which may be part of the molecule
153. Several Authors (1985) Polyurethanes In: Oertel G (ed) Plastics Handbook, Carl Hanser, München vol 7 p 117–590
154. *Woods G* (1987) The ICI Polyurethanes Book, Wiley, Chichester
155. The Society of the Plastics Industry, Polyurethane Division 355 Lexington Ave New York 10017 USA: Proceedings of past Technical and Marketing Conferences
156. Proceedings of Cellular and non-Cellular Polyurethanes International Conference June 9–13, 1980 Strasbourg France, Carl Hanser, München and: International Polyurethane Congress

September 29 – October 2, 1987, Aachen Federal Republic of Germany, Technomic Publishing, Lancaster; both conferences were jointly sponsored by: The Society of the Plastics Industry (SPI) Polyurethane Division, 355 Lexington Ave, New York NY 10017, USA and Fachverband Schaumkunststoffe e.V. (FSK) 6000 Frankfurt, Federal Republic of Germany

157. Conte A, Cossi G (1981) J Chromatogr 213: 162
158. DIN Safety Data Sheet 028650/04 Desmodur^R L 75 (July 1988) Bayer AG Business Group LS
159. DIN Safety Data Sheet 028596/04 Desmodur^R HL (August 1988) Bayer AG Business Group LS
160. DIN Safety Data Sheet 028677/05 Desmodur^R N 3200 (April 1988) Bayer AG Business Group LS
161. DIN Safety Data Sheet 028731/07 Desmodur^R N 75 (August 1988) Bayer AG Business Group LS
162. DIN Safety Data Sheet 028987/06 Desmodur^R Z 4370 (July 1988) Bayer AG Business Group LS
163. DIN Safety Data Sheet 028715/08 Desmodur^R N 3390 (June 1988) Bayer AG Business Group LS
164. Potapova M P, Lushchik V, Ermolaeva T A, Pronina I A (1974) Lakrokas Mat Ikh Primen 4: 73
165. Potapova M P, Ermolaeva T A (1978) Metod Anal Kontr Khim Prom-Sti 7: 3
166. Wicks Z W (1975) Prog Org Coat 3: 73
167. DIN Safety Data Sheet 029835/04 Desmodur^R E 21 (February 1988) Bayer AG Business Group LS
168. Dhein R, Kreuder H J, Rudolph H (1972) Farbe und Lack 78: 1060
169. Brushwell W (1975) Farbe und Lack 81: 33
170. Brushwell W (1976) Farbe und Lack 82: 134
171. DIN Safety Data Sheet 072358/06 Desmodur^R R (August 1988) Bayer AG Business Group Rubber
172. DIN Safety Data Sheet 072331/06 Desmodur^R RF (August 1988) Bayer AG Business Group Rubber
173. Skeist J (1976) Handbook of Adhesives, 2nd edn Van Nostrand Reinhold, New York
174. Deppe H J, Ernst K (1971) Holz Roh Werkst 29: 45
175. Kögler H, Scholich K (1978) Gießerei 65: 101
176. Zapp J A (1957) A M A Arch Ind Health 15: 324
177. Henschler D, Assmann W (1962) Arch f Toxikol 19: 364
178. Duncan B, Scheel L D, Fairchild E J, Killens R, Graham S (1962) Am Ind Hyg Ass J 23: 447
179. Bunge W, Ehrlicher H, Kimmerle G (1977) Z Arbeitsmed Arbeitsschutz Prophyl 4 (special ed): 1
180. Adams W G F, Carney I F, Chamberlain J D, Paddle G M (1978) Lancet 1: 1308
181. Carney I F (1980) Eur J Cell Plast July 3: 78
182. Rampy L W, Löser E, Lyon J P, Carney I (1983) Proc SPI 6th Int Techn Conf p 413
183. Karol M H (1986) CRC Critical Rev Toxicol 16: 349
184. Thorne P S, Hillebrand J S, Lewis G R, Karol M H (1987) Toxicol Appl Pharmacol 87: 155
185. INRS (1987) Fiche Toxicologique No. 46 Diisocyanate de toluylène
186. Woolrich P F, Rye W A (1969) J Occup Med 11: 184
187. INRS (1987) Fiche Toxicologique No. 129 4,4'-Diisocyanate de diphénylméthane
188. Smyth H F, Carpenter C P, Weil C S, Pozzani U C, Striegel J A, Nycum J S (1969) Am Ind Hyg Assoc J 30: 470
189. INRS (1988) Fiche Toxicologique No. 164 1,6-Diisocyanate d'hexaméthylène
190. Takeda Chemical Industries Ltd, Osaka, Japan (1988) personal communication
191. Documentation of the TLV and Biological Exposure Indices (1986) (5th edn) Cincinnatti, American Conference of Governmental Industrial Hygienists, p 392
192. INRS (1987) Fiche toxicologique no. 166 diisocyanate d'isophorone
193. Deutsche Forschungsgemeinschaft (1973) Senatskommission zur Prüfung gesundheitsschädlicher Arbeitsstoffe, Isophorondiisocyanat, MAK-Wert-Begründung
194. Schafer E W, Bowles W A, Hurblut J (1983) Arch Envir Contam Toxicol 12: 355
195. Harton E E, Rawl R R (1976) US Department of Commerce NTIS Publication 265975
196. Gordon T, Sheppard D, McDonald D M, Distefano S, Scypinsky L (1985) Am Rev Respir Dis 132: 1106
197. Gordon T, Milligan S A, Levin J, Thompson J E, Fine J M, Sheppard D (1987) Am Rev Respir Dis 135: 854

198. *Sangha G K, Alarie Y C* (1979) Toxicol Appl Pharmacol 50: 533
199. *Weyel D A, Rodney B S, Alarie Y C* (1982) Toxicol Appl Pharmacol 64: 423
200. *Frank H G* (1964) Zahnmed Inaug Diss, Würzburg
201. Deutsche Forschungsgemeinschaft (1971) Senatskommission zur Prüfung gesundheitsschädlicher Arbeitsstoffe, Hexamethylendiisocyanat, MAK-Wert-Begründung
202. Material Safety Data Sheet Dicyclohexylmethane-4,4′-Diisocyanate Sept (1988) Mobay Chemical Corporation Pittsburgh PA USA
203. E.I. Dupont de Nemours and Company, Haskell Laboratories (1978) Unpublished results
204. *Sangha G K, Matijak M, Alarie Y C* (1981) Toxicol Appl Pharmacol 57: 241
205. *Valentini J E, Wong K L, Alarie Y C* (1983) Toxicol Appl Pharmacol 69: 461
206. *Niewenhuis R, Scheel L D, Stemmer K, Killens R* (1964) Am Ind Hyg Ass J 26: 143
207. *Buckley L A, Jiang X Z, James K T, Morgan K T, Barrow C S* (1984) Toxicol Appl Pharmacol 74: 417
208. *Stevens M A, Palmer R* (1970) Proc R Soc Med 63: 22
209. *Collins C, Bier C, Freidman M, Breckenridge C, Walker J, Cornell S* (1984) Toxicologist 4: 160
210. *Karol M H* (1983) Toxicol Appl Pharmacol 68: 229
211. *Karol M H, Stadler J C* (1985) Toxicologist 5: 6
212. *Wong K L, Karol M H, Alarie Y C* (1985) J Toxicol Env Health 15: 137
213. E.I. Dupont de Nemours and Company Haskell Laboratories (1984) FYIFYIOTS-0584-0303 (Office of Toxic Substances Environmental Protection Agency) Unpublished reports May 12, 1965; May 20, 1971; April 16, 1974; Wilmington DE USA
214. International Isocyanate Institute (1984) Unpublished information
215. *Weyel D A, Schaffer R B* (1985) Toxicol Appl Pharmacol 77: 427
216. *Bunge W* (1975) Med Inaug Diss, Würzburg
217. *Lomonova G V, Frolova I N* (1968) Gig Trud Prof Zabol 12: 40
218. *Schütze T* (1971) Med Inaug Diss, Würzburg
219. *Kohler R* (1970) Med Inaug Diss, Würzburg
220. Tests performed by the Fraunhofer-Institut für Toxikologie und Aerosolforschung, D 5948 Schmallenberg; reports dated April 2, 1981 (unpublished);
OECD Guideline for Testing of Chemicals: "Acute Dermal Irritancy/Corrosivity", Adopted 404;
OECD Guideline for Testing of Chemicals: "Acute Eye Irritancy/Corrosivity", Adopted 405.
221. *Duprat P, Gradiski D, Marignac B* (1976) Eur J Toxicol 9: 41
222. *Peschel J* (1970) Derm Mschr 156: 691
223. *Karol M H, Hauth B A, Riley E J, Magreni C M* (1981) Toxicol Appl Pharmacol 58: 221
224. *Koschier F J, Burden E J, Brunkhorst C S, Friedman M A* (1983) Toxicol Appl Pharmacol 67: 401
225. *Tanaka K I* (1979) Japn J Ind Health 21: 456
226. *Tominaga M, Kohno S, Tanaka K, Ohata K* (1985) Japn J Pharmacol 1985: 163
227. *Tanaka K, Takeoka A, Hanada S, Okamoto Y, Ino T, Okuizumi J* (1985) Japn J Allergol 34: 128
228. *Ohsawa T, Ishizu S* (1983) J Tokyo Women Med Coll 53: 273
229. *Ohsawa T* (1983) J Tokyo Women Med Coll 53: 237
230. *Doe J E, Hicks R, Milburn G M* (1982) Int Arch Allergy Appl Immunol 68: 275
231. *Chadwick D H, Cleveland T H* (1981) In: Kirk-Othmer Encyclopedia of Chemical Technology, 3rd edn, Wiley, New York, vol 13 p 789
232. *Tanaka K I, Takeoka A, Nishimura F, Hanada S* (1987) Contact Dermat 17: 199
233. *Kondratev G G, Mustaev R K* (1969) Gig Primen Polim Mater Izdelii Nikh 1: 290
234. *Karol M H, Magreni C* (1982) Toxicol Appl Pharmacol 65: 291
235. *Stadler J C, Karol M H* (1985) Toxicol Appl Pharmacol 78: 445
236. INRS (1987) Fiche Toxicologique No. 166 Diisocyanate d'isophorone
237. *Karol M H, Hansen G A, Brown* (1984) *Fundam Appl Toxicol* 4: 284
238. *Brown W E, Green A H, Karol M H, Alarie Y C* (1982) Toxicologist 2: 164
239. *Dewair M, Baur X, Mauermayer R* (1983) Int Arch Occup Environ Health 52: 257
240. *Brown W E, Shamoo A Y, Hill B L, Karol M H* (1984) Toxicol Appl Pharmacol 73: 105
241. *McKay R T, Brooks S M* (1983) Am Rev Respir Dis 127: 174
242. *McKay R T, Brooks S M* (1983) Am Rev Respir Dis 128: 50
243. *McKay R T, Brooks S M* (1984) Am Rev Respir Dis 129: 296
244. *Karol M H, Stadler J C, Underhill D, Alarie Y C* (1981) Toxicol Appl Pharmacol 61: 277
245. *Stadler J C, Karol M H* (1982) Toxicol Appl Pharmacol 65: 323

246. *Karol M H, Stadler J C, Magreni C* (1985) Fund Appl Toxicol 5: 459
247. *Van Ert M, Battigelli M C* (1975) Annals of Allergy 35: 142
248. *Karol M H, Ioset H H, Alarie Y C* (1978) Am Ind Hyg Ass J 39: 546
249. *Karol M H* (1980) Am Rev Respir Dis 112: 965
250. *Karol M H, Hauth B A* (1982) Fundam Appl Toxicol 2: 108
251. *Tanaka K, Kawai M, Maekawa N* (1982) Jpn J Allergol 31: 1004
252. *Tse C S T, Chen E, Bernstein I L* (1979) Am Rev Respir Dis 120: 829
253. *Chen S E, Bernstein I L* (1982) J Allergy Clin Immunol 70: 383
254. *Mullin L S, Wood C K, Krivanek N D* (1983) Toxicol Appl Pharmacol 71: 113
255. *Bernstein I L, Splansky G L, Chen S E, Vinegar A* (1982) J Allergy Clin Immunol 70: 393
256. *Patterson R, Harris K E, Pruzansky J J, Zeiss CR* (1982) J Lab Clin Med 99: 615
257. *Patterson R, Zeiss C R, Harris K E* (1983) J Allergy Clin Immunol 71: 604
258. *Tanaka K, Okamoto Y, Takeoka A, Ino T* (1984) Japn J Allergol 33: 199
259. *Kido T, Yamada Y, Teranishi H* (1983) Japn J Allergol 32: 131
260. *Cibulas W, Murlas C G, Miller M L, Vinegar A, Schmidt D J, McKay R T, Bernstein I L, Brooks S M* (1986) J Allergy Clin Immunol 77: 828
261. *Karol M H* (1985) In: J Dean et al (eds), Immunotoxicology and Immunopharmacology. Raven, New York, p 475
262. *Karol M H* (1985) In: Hodgson E, Bernd J, Philpot R M (eds), Reviews in Biochemical Toxicology Elsevier, New York, vol 7 p 105
263. E.I. Dupont de Nemours and Company Haskell Laboratories (1976) Report MR 2356 (HL 784-76) Wilmington USA
264. *Anderson D, Styles J A* (1978) Appendix II Br J Can 37: 924
265. *Andersen M, Brindrup M, Kiel P, Larsen H, Maxild J* (1980) Scand J Work Envir Health 6: 221
266. *Diller W F* (1981) Scand J Work Envir Health 7: 237
267. *Andersen M, Binderup M L, Kiel P, Larsen H, Maxild J* (1982) Scand J Work Envir Health 8: 80
268. *Sopach E D, Boltromeyuk L P* (1974) Gig Sanit 7: 10
269. National Toxicology Programme (1983) NIH publication no. 83-2507
270. *Styles J A* (1978) Appendix III Br J Can 37: 931
271. *Löser E* (1983) Toxicol Lett 15: 71
272. *Maki-Paakkanen J, Paakkanen J M, Norppa H* (1987) Toxicol Lett 36: 37
273. *Shimizu H, Suzuki Y, Takemura N, Goto S, Matsushita H* (1985) Japan J Ind Health 27: 400
274. E.I. Dupont de Nemours and Company (1976) Haskell Laboratories Report MR 2389 (HL 463-76) Wilmington USA
275. *Ader A W, Carney I F, Löser E* (1987) Proc FSK/SPI International Polyurethane Congress Aachen p 188
276. National Toxicology Program (August 1986) Technical Report Series No. 251 US Department of Health Service National Institute of Health
277. World Health Organization IARC Monograph (1978) 2,4-Toluene diamine, WHO, Geneva, vol 16 p 83
278. World Health Organization IARC Monographs (1982) 2,4-Toluene diamine, WHO, Geneva, Suppl. 4 vol 1–29 p 268
279. *Duff P B* (1983) Proc SPI 6th international technical/marketing conference p 408
280. *Brochhagen F K, Grieveson B M* (1984) Cell Polym 3: 11
281. *Gilbert D S* (1987) Proc FSK/SPI international polyurethane conference, p 162
282. *Brown S L, Chan F Y, Jones J L* (1975) Research Program on Hazard Priority Ranking of Manufactured Chemicals Phase II Final Report Chemicals 21–41, Washington DC: National Science Foundation p 24-B-1
283. *Domsch K H, Martens R* (1981) Water Air Soil Poll 15: 503
284. *Curtis M W, Copeland T L, Ward C H* (1979) Water Research 13: 137
285. *Bengtsson B E, Tarkpea M* (1983) Mar Poll Bull 14: 213
286. *Tu D, Fetsch R* (1980) Forschungsbericht 104.08.144, Universität Stuttgart
287. *Grieveson B M, Reeve B* (1983) Cell Polym 2: 165
288. *Rando R J, Abdel-Kader H M, Hammad Y Y* (1984) Am Ind Hyg Ass J 45: 199
289. *Sporon-Fiedler A* (1986) Cell Polym 5: 369
290. *Creyf H, Hurd R, Powell D* (1986) Cell Polym 5: 461
291. *Dyson W L, Hermann E R* (1971) Am Ind Hyg Ass J 32: 741
292. *Holland D G, Rooney T A* (1977) J Occup Med 19: 239
293. *Goellner K* (1978) Master's Thesis, Univ. Pittsburgh, USA

294. *Dharmarajan V, Weill H, Self C W* (1978) Am Ind Hyg Ass J 39: 414
295. *Holdren M W, Spicer C W, Riggin R M* (1984) Am Ind Hyg Ass J 45: 626
296. *Duff P B* (1984) Proc SPI 29th SPI Techn/Mark Conf, p 9
297. *Walker R F, Pinches M A* (1981) Am Ind Hyg Ass J 42: 392
298. *Sandridge R L* (1982) Am Ind Hyg Ass J 43: A-16
299. *Walker R F* (1982) Am Ind Hyg Ass J 43: A-17
300. DFG Deutsche Forschungsgemeinschaft (1988) Mitteilung XXIV der Senatskommission zur Prüfung gesundheitsschädlicher Arbeitsstoffe, VCH Verlagsgesellschaft, Weinheim
301. Quoted from DFG Deutsche Forschungsgemeinschaft (1985) Report XXI of the Commission for the Investigation of Health Hazards of Chemical Compounds in the Work Area, VCH Verlagsgesellschaft, Weinheim
302. American Conference of Governmental Industrial Hygienists (1988) Threshold Limit Values and Biological Exposure Indices for 1988–1989, Cincinnati, OH, 45211-4438
303. Circulaire du Ministère du Travail du 5 Mai (1986)
304. Occupational Exposure Limits 1985 Guidance Note EH 40/85 from the Health and Safety Executive
305. *Diller W F, Brochhagen F K, Slawyk W* (1985) Zbl Arbeitsmed 35: 88
306. *Nielsen J, Sangoe C, Winroth G, Hallberg T, Skerfving S* (1985) Scand J Work Envir Health 11: 51
307. *Brochhagen F K* (1985) Scand J Work Envir Health 11: 497
308. Courtesy of the International Isocyanate Institute Inc. More information on the origin of the various data can be obtained from the I.I.I. Scientific Office, P.O. Box 42, Hexagon House, Blackley, Manchester, M9 3DA, England
309. 29 CFR 1910.134 OSHA General Industry Standard on Respiratory Protection
310. *Brochhagen F K, Hauptmann G* (1985) Zbl Arbeitsmed 35: 70
311. *Smith D B, Henderson R* (1975) J Occup Med 17: 413
312. *Pagnotto L D, Wegman D H, Henderson R* (1976) J Occup Med 15: 523
313. *Walber U* (1985) Polyurethanes. In: Oertel G (ed), Plastics Handbook. Carl Hanser, München, vol 7 p 593
314. *Glazemakers J* (1986) Cah Med Trav 23: 213
315. Technical Information 5 (1982) International Isocyanate Institute Inc
316. *Keller J* (1985) Zbl Arbeitsmed 35: 74
317. *Brenner K S* (1987) Proc SPI/FSK Polyurethane World Congress, Aachen p 156
318. *Booth K S, Dharmarajan V, Lingg R D, Darr W C* (1983) Proc SPI 28th Techn Mark Conf
319. *Rietz B* (1983) Dansk Kemi 4: 112
320. *Omae K, Sakurai H* (1984) Chemical Factory (Japan) 28: 1441
321. *Groves J A, Brown R H* (1985) Urethanes Technol Dec 85: 35
322. *Brown R H, Ellwood P A, Groves J A, Robertson S M* (1987) Cell Pol 6: 1
323. *Langhorst M L, Coyne L B* (1987) Anal Chem 59: 1R
324. *Keller J, Sandridge R L* (1979) Anal Chem 51: 1868
325. *Tucker S P, Arnold J E* (1982) Anal Chem 54: 1137
326. *Westerhoff C B, Ozkan M, Sowinski E J* (1983) J Elastomers Plast 15: 3
327. *Andersson K, Gudehn A, Levin J O, Nilsson C A* (1983) Am Ing Hyg Ass J 44: 802
328. *Rosenberg C, Tuomi T* (1984) Am Ind Hyg Ass J 45: 117
329. *Dharmarajan V* (1979) J Env Pathol Toxicol 2: 1
330. *Marcali K* (1957) Anal Chem 29: 552
331. *Skonieczny R F* (1963) Am Ind Hyg Ass J 24: 17
332. *Grim K E, Linch AL* (1964) Am Ind Hyg Ass J 25: 285
333. *Pilz W* (1965) Mikrochim Acta 4: 687
334. *Meddle D W, Radford W D, Wood R* (1969) Analyst 94: 369
335. *Miller A J, Mueller F X* (1975) Am Ind Hyg Ass J 36: 477
336. *Mushkin Y I, Smirnova N F, Gainutdinova K M* (1976) Tr Giap 42: 49
337. *Rubio J V* (1982) Medicina y Seguridad del Trabajo 30: 119
338. *Pilz W* (1970) Mikrochim Acta 27: 504
339. *Rando R J, Hammad Y Y* (1985) Am Ind Hyg Ass J 46: 206
340. *Meddle D W, Wood R* (1970) Analyst 95: 402
341. *Vogel J, Keller J* (1987) Proc SPI/FSK Polyurethane World Congress, Aachen p 162
342. *von Eicken S* (1959) Z Anal Chem 171: 136
343. *Pilz W, Johann I* (1969) Z Anal Chem 248: 149

344. *Pilz W, Johann I* (1970) Microchim Acta 27: 351
345. *Walker R F, Pinches M A* (1979) Analyst 104: 928
346. *Wheals B B, Thomson J* (1967) Chem & Ind (May 6): 753
347. *Schanche G W, Hermann E R* (1974) Am Ind Hyg Ass J 35: 47
348. *Skarping G, Smith B E F, Dalene M* (1985) J Chromatogr 331: 331
349. *Holland D G, Rooney T A* (1977) J Occup Med 19: 238
350. *Skarping G, Sangoe C, Smith B E F* (1981) J Chromatogr 208: 313
351. *Skarping G, Renman L, Sangoe C, Mathiasson L, Dalene M* (1985) J Chromatogr 346: 191
352. *Fukabori S, Nakaaki K* (1986) J Sci Labour 62: 591
353. *Bishop R W, Ayers T A, Esposito G G* (1983) Am Ind Hyg Ass J 44: 151
354. *Costarca L, Moraru A, Bacaloglu R* (1984) Buletinul Stiintific si Technical Inst Politech 'Traian Vuia' Timisoara 29: 77
355. *Audunsson G, Mathiasson L* (1983) J Chromatogr 261: 253
356. *Audunsson G, Dalene M, Joensson J A, Loevkvist P, Mathiasson L, Skarping G* (1985) Int J Env Anal Chem 20: 85
357. *Keller J, Dunlap K L, Sandridge R L* (1974) Anal Chem 46: 1845
358. *Ellwood P A, Hardy H L, Walker R F* (1981) Analyst 106: 85
359. *Dunlap K L, Sandridge R L, Keller J* (1976) Anal Chem 48: 497
360. *Bagon D A, Purnell C J* (1980) J Chromatogr 190: 175
361. *Sangoe C* (1979) J Liq Chromatogr 2: 763
362. *Hakes D C, Johnson G D, Marhevka J S* (1986) Am Ind Hyg Ass J 47: 181
363. *Graham J D* (1980) J Chromatogr Sci 18: 384
364. *Rosenberg C* (1984) Analyst 109: 859
365. *Wu W S, Huang L K, Gaind V S* (1986) Am Ind Hyg Ass J 47: 482
366. *Hardy H L, Walker R F* (1979) Analyst 104: 890
367. *Chang S N, Burg W R* (1982) J Chromatogr 246: 113
368. *Goldberg P A, Walker R F, Ellwood P A, Hardy H L* (1981) J Chromatogr 212: 93
369. *Nakaaki K, Koike S, Takada Y* (1987) J Sci Labour 63: 1
370. *Nieminen E H, Saarinen L H, Laakso J T* (1983) J Liq Chromatogr 6: 453
371. *Lipski K* (1982) Ann Occup Hyg 25: 1
372. *Shoemaker R A* (1981) J Chromatogr Sci 19: 321
373. *Levine S P, Hoggatt J H, Chladec E, Jungclaus G, Gerlock J L* (1979) Anal Chem 51: 1106
374. *Kormos L H, Sandridge R L, Keller J* (1981) Anal Chem 53: 1122
375. *Sangoe C, Zimerson E* (1980) J Liq Chromatogr 3: 971
376. *Andersson K, Gudehn A, Hallgren C, Levin J O, Nilsson C A* (1983) Scand J Work Environ Health 9: 497
377. *Rietz B* (1985) Anal Letter 18: 1193
378. *Warwick C J, Bagon D A, Purnell C J* (1981) Analyst 106: 676
379. *Meyer S D, Tallmann D E* (1983) Anal Chim Acta 146: 227
380. *De Pascale A, Cobelli L, Paladino R, Pastorello L, Frigerio A, Sala C* (1983) J Chromatogr 256: 352
381. *Reilly D A* (1968) Analyst 93: 178
382. MDA Scientific, Glenview Illinois USA, Literature on isocyanate monitors
383. SKC Ltd., Poole, Dorset England
384. *Dharmarajan V, Rando R J* (1980) Am Ind Hyg Ass J 41: 869
385. *Rosenberg C, Pfäffli P* (1982) Am Ind Hyg Ass J 43: 160
386. *Nutt A* (1983) Anal Proc 20: 63
387. *Sciarra G, Innocenti A, Bozzi N* (1984) Med Lav 75: 491
388. *McMahon R* (1983) Proc SPI 28th Techn Mark Conf
389. *Mazur G, Baur X, Pfaller A, Roemmelt H* (1986) Int Arch Occup Environ Health 58: 269
390. *Rando R J, Duvoisin P F, Abdel-Kader H, Hammad Y Y* (1987) Am Ind Hyg Ass J 48: 574
391. *Alder J F, Isaac C A* (1981) Anal Chim Acta 129: 163
392. *Alder J F, Isaac C A* (1981) Anal Chim Acta 129: 175
393. *Fielden P R, McCallum J J, Stanios T, Alder J F* (1984) Anal Chim Acta 162: 85
394. *Morrison R C, Guilbault G G* (1985) Anal Chem 57: 2342
395. Graseby Dynamics Ltd., Watford Herts England
396. *Rosenberg C, Savolainen H* (1985) J Chromatogr 323: 429
397. *Rosenberg C, Savolainen H* (1986) J Chromatogr 358: 385

398. Project Report No 7 (1985) Research Triangle Institute 2227, 00–06 P, submitted to National Institute of Environmental Health Science
399. *Rosenberg C, Savolainen H* (1986) Analyst 111: 1069
400. Centaur Management Consultants, Inc. (1976) Technological Feasibility Assessment and Inflationary Impact Analysis for MDI and TDI, final report 3-126
401. National Institute for Occupational Safety and Health (1978) DHEW (NIOSH) Publication No. 78-215, p 1
402. The 'Deutsche Forschungsgemeinschaft' has composed under the heading 'Stäube und Rauche (feste Schwebstoffe)' – Dust(s) and Smoke(s) {Solid Suspended Particles} – a detailed definition which should be applicable to the various types of particles bearing free -NCO-groups; see 'Mitteilung XXIV der Senatskommission zur Prüfung gesundheitsschädlicher Arbeitsstoffe' (1988) VCH Verlagsgesellschaft, Weinheim
403. *Porter C V, Higgins R L, Scheel L D* (1975) Am Ind Hyg Ass J 36: 159
404. *Diem J E, Jones R N, Hendrick D J, Glindmeyer H W, Butcher B T, Salvaggio J E, Dharmarajan D, Weill H* (1982) Am Rev Respir Dis 126: 420, see also Robins J, Greaves I A, Eisen E A (1983) Am Rev Respir Dis 128: 327
405. *Peters J M, Murphy R L H, Pagnotto L D, van Ganse W F* (1968) Arch Environ Health 16: 642
406. *Peters J M, Murphy R L H, Ferris B G* (1969) Brit J Ind Med 26: 115
407. *Gee J B, Morgan W K C* (1985) J Occup Med 27: 15
408. *Kelly R M, Egilman D, Gordon J, Gee J B L, Morgan W K C* (1985) J Occup Med 27: 469
409. *Musk A W, Peters J M, DiBerardinis L, Murphy R L H* (1982) J Occup Med 24: 746
410. *Rando R J, Abdel-Kader H, Hughes J, Hammad Y Y* (1987) Am Ind Hyg Ass J 48: 580
411. *Cavelier C, Pham Q T, Merlau P, Mur J M, Rombach F, Cicolella A, Bui Dinh Long L* (1977) Cah Notes Doc 88: 315
412. *Diller W F, Nießen J, Klebert W* (1985) Zbl Arbeitsmed 3: 85
413. *Diller W F, Herbert E* (1982) Zbl Arbeitsmed 32: 128
414. *Diller W F, Alt E, Klebert W* (1985) Zbl Arbeitsmed 3: 82
415. Deutsche Forschungsgemeinschaft (1984) Senatskommission zur Prüfung gesundheits-schädlicher Arbeitsstoffe Toxikologisch-arbeitsmedizinische Begründung von MAK-Werten, Diisocyanate, Verlag Chemie, Weinheim
416. *Berlin M, Franke U* (1981) Verkehrsmedizin 28: 151
417. *McKerrow C B, Davies H J, Jones A P* (1970) Proc Roy Soc Med 63: 376
418. *Hussman P* (1973) Pracov Lék 25: 242
419. *Saia B, Rossi A, Mastrangelo G* (1973) Med Lavoro 64: 143
420. *Saia B, Fabbri L, Mapp C, Marcer G, Mastrangelo G* (1976) Med Lavoro 67: 278
421. *Fabbri L, Saia B, Mapp C, Marcer G, Mastrangelo G* (1976) Med Lavoro 67: 305
422. *Mapp C, Fabbri L, Marcer G, Mastrangelo G, Saia B* (1976) Lavoro Umano 28: 17
423. *Alexandersson R, Kolmodin-Hedman, Hedenstierna G, Magnusson M* (1980) Arbete och Hälsa 5: 1
424. *Oleru U G* (1980) Am Ind Hyg Ass J 41: 595, see also Oleru UG (1980) Environ Res 23: 137
425. *Bugler J* (1982) Diss, Univ Manchester, England
426. *Mur J M, Meyer-Bisch C, Cavelier C, Pham Q T, Lacube P* (1982) Cah Notes Doc 106: 83
427. *Pham Q T, Meyer-Bisch C, Gaertner M, Mur J M, Pierre F, Huez D* (1982) Arch Mal Prof 43: 97
428. *White W G, Sugden E, Morris M J, Zapata E* (1980) Lancet I: 756
429. *Axford A T, McKerrow C B, Jones A P, Le Quesne P M* (1976) Br J Ind Med 33: 65
430. *Le Quesne P M, Axford A T, McKerrow C B, Jones A P* (1976) Br J Ind Med 33: 72
431. *Candura F, Moscato G* (1984) Brit J Ind Med 41: 552
432. *Brochhagen F K* (1985) Br J Ind Med 42: 567
433. *Bernoulli R G, Engelhart G, Velvart J* (1978) Schweiz Med Wschr 108: 866
434. *Schmauss A K, Schreiber F U, Ziegler P F, Brzyk I* (1982) Med Akt 8: 108
435. *Diller W F* (1985) Zbl Arbeitsmed 35: 66
436. *Ehrlicher H* (1974) Pneumologie 150: 155
437. *Hama G M* (1957) Am Med Assoc Arch Ind Health 16: 232
438. *Walworth H T, Virchow W E* (1959) Am Ind Hyg Assoc J 20: 205
439. *Munn A* (1960) Trans Assoc Ind Med Off 9: 134
440. *Seidel H, Pohle H* (1960) Tuberk Arzt 14: 675
441. *Brugsch H G, Elkins H B* (1963) New England J Med 268: 353
442. *Munn A* (1968) Adv Polyurethane Technol 16: 299
443. *Reinl W, Schnellbächer F* (1974) Zbl Arbeitsmed 24: 106

444. *Luckenbach M, Kielar R* (1980) Am J Ophthalm 90: 682
445. *Paggiaro P L, Rossi E, Lastrucci L, Pardi F, Pezzini A, Buschieri L* (1985) J Occup Med 27: 51
446. *Reinl W* (1953) Zbl Arbeitsmed Arbeitssch 3: 103
447. *Lineweaver P G* (1972) J Occup Med 14: 25
448. *Gervais P, Diamant-Berger O, Roux Y, Niel E* (1973) J Eur Toxicol 6: 320
449. *Fabbri L, Danieli D, Meli S, Bevilacqua P, Crescioli S, Mapp C* (1987) Bull Eur Physiopathol Respir 23: 352
450. US Department of Health and Human Services, National Institute for Occupational Safety and Health (1978) Pocket Guide to Chemical Hazards
451. *Williamson K S* (1964) Trans Assoc Ind Med Officers 14: 81
452. *Adams W G F* (1970) Proc R Soc Med 63: 20
453. *Adams W G F* (1970) Proc R Soc Med 63: 378
454. *Adams W G F* (1975) Brit J Ind Med 32: 72
455. *Adams W G F, Carney I F, Chamberlain J D, Paddle G M* (1978) Lancet 17 June: 1308, see also Peters J M, Wegman D H (1978) Lancet 26 Aug: 472
456. *Ehrlicher H, Brochhagen F K* (1976) Proc Plastics Rubber Institute Conf
457. *Omae K, Sakurai H, Toyama T, Nakadate T, Higashi T, Tsugane S, Aizawa Y* (1983) Scand J Work Env Health 10: 137
458. *Omae K* (1984) Int Arch Occup Env Health 55: 1
459. *Pitzalis G* (1980) Diss, Univ Pavia
460. *Pitzalis G, Colle B, Orefice U* (1980) Proc Congr Naz Med Lav Sorrent
461. *Catenacci G, Franco G, Pozzoli L* (1972) Lavoro Umano 24: 73
462. *Franzinelli A, Mariotti F, Innocenti A* (1978) Med Lavoro 69: 163
463. *Kolmodin.Hedman B, Alexandersson R, Hedenstierna G* (1980) Arbete och Hälsa 10: 4
464. *Peters J M, Murphy R L H, Pagnotto L D, Whittenberger J L* (1970) Arch Environ Health 20: 364
465. *Wegman D H, Pagnotto L D, Fine L D, Peters J M* (1974) J Occup Med 16: 258
466. *Wegman D H, Peters J M, Pagnotto L D, Fine L J* (1977) Brit J Ind Med 34: 196
467. Deutsche Forschungsgemeinschaft (1972) Toxikologisch-arbeitsmedizinische Begründung von MAK-Werten, Toluylendiisocyanate, Verlag Chemie, Weinheim
468. *Hill R N* (1970) Proc R Soc Med 63: 375
469. *Weill H B, Butcher B, Diem J E, Dharmarajan V, Glindmeyer H, Jones R, Carr J, O'Neil C, Salvaggio J* (1979) NIOSH final report 210-75-006, New Orleans, U.S.A.
470. *Peters J M, Wegman D H* (1975) Env Health Persp 11: 97
471. *Wegman D H, Main D M, Musk A W* (1981) Am Rev Respir Dis 123: 145
472. *Wegman D H, Musk A W, Main D M, Pagnotto L D* (1982) Am J Ind Med 3: 209
473. *Bruckner H C, Avery S B, Stetson D M, Dodson V N* (1968) Arch Environ Health 16: 619
474. *Karol M H* (1981) J Occup Med 23: 741
475. *Butcher B T, Salvaggio J E, O'Neil C E, Weill H, Garg O* (1977) J Allergy Clin Immunol 59: 223
476. *Butcher B T, Karr R M, O'Neil M S, Wilson M R, Dharmarajan V, Salvaggio J E, Weill H* (1979) J Allergy Clin Immunol 64: 146
477. *Butcher B T, O'Neil C E, Reed M A, Salvaggio J E, Weill H* (1982) J Allergy Clin Immunol 70: 231
478. *Baur X, Dewair M, Fruhmann G* (1984) J Allergy Clin Immunol 73: 610
479. *Gallagher J S, Tse C S T, Pharm D, Brooks S M, Bernstein I L* (1981) J Occup Med 23: 610
480. *Karol M H, Sandberg T, Riley E J, Alarie Y C* (1979) J Occup Med 21: 354
481. *Karol M H, Riley E J, Alarie Y C* (1979) J Env Health Sci 13: 221
482. *Game C J A* (1982) Am Ind Hyg Ass J 43: 759
483. *Kido T* (1983) Allergy 32: 131
484. *van Ert M, Battigelli M C* (1975) Ann Allergy 35: 142
485. *Davies R J* (1978) J Allergy Clin Immunol 60: 223
486. *McKay R T, Brooks S M, Johnson G* (1981) Chest (Suppl) 80: 61
487. *Danks J M, Cromwell O, Buckingham J A, Newman-Taylor A J, Davies R J* (1981) Clin Allergy 11: 161
488. *Bernstein J* (1982) J Allergy Clin Immunol 70: 24
489. *Thurmann G B, Simms B S, Goldstein A L, Kilian D J* (1978) Toxicol Appl Pharmacol 44: 617
490. *Markham T N, Fishburn C W* (1967) J Occup Med 9: 41
491. *Smith A B, Brooks S M, Blanchard J, Bernstein I L, Gallagher J* (1980) J Occup Med 22: 327
492. *O'Brien I M, Newman-Taylor A J, Burge P S, Harries M G* (1979) Clin Allergy 9: 7

493. *Gallagher J D et al* (1982) Am Rev Respir Dis 125: 167
494. *Innocenti A, Franzinelli A, Sartorelli E* (1981) Med Lavoro 72: 231
495. *Friedman S A* (1982) Am Rev Respir Dis 125: 167
496. *Blake B L, McKay J B, Rainey H B, Weston W J* (1965) J Coll Radiol Aust 9: 45
497. *Malo J L, Ouimet G, Cartier A, Levitz D, Zeiss CR* (1983) J Allergy Clin Immunol 72: 413
498. *Fink J N, Schlueter D P* (1978) Am Rev Respir Dis 118: 956
499. *Lob M, Boillat M A* (1981) Schweiz Med Wschr 111: 150
500. *Charles J, Bernstein A, Jones B, Edwards J H, Seal R M E, Seaton A* (1976) Thorax 31: 127
501. *Zeiss C R, Kanellakes T M, Bellone J D, Levitz D, Pruzansky J J, Patterson R* (1980) J Allergy Clin Immunol 65: 346
502. *Baur X, Fruhmann G* (1983) Verh Jahrestagung Dtsch Ges Arbeitsmed, Göttingen, p 81
503. *Sartorelli E, Catalano P, Franzinelli A, Innocenti A, Severi A* (1976) 39 Congr Med Lav, Fiuggi Terme
504. *Taylor G* (1970) Proc R Soc Med 63: 379
505. *Rakow A B, Baier E J* (1971) HSMAH Health Rep 86: 663
506. *Pepys J, Pickering C A C, Breslin A B X, Terry D J* (1972) Clin Allergy 2: 225
507. *Elkins H B, McCare G W, Brugsch H G, Fahy J P* (1962) Am Ind Hyg Ass J 23: 265
508. *Vandervort R, Shama G* (1973) NIOSH rep TR 090-74, Cincinnatti, U.S.A.
509. *Diller W F, Alt E, Baur X, Fruhmann G* (1980) Zbl Arbeitsmed 30: 100
510. *Chester E H, Martinez-Catinchi F L, Schwartz H J, Horowitz J, Fleming G M, Gerblich A A, McDonald G W, Grethawer R* (1979) Chest 75: 229
511. *Zammit-Tabona M, Sherkin M, Kijek K, Chan H, Chan-Yeung M* (1983) Am Rev Respir Dis 128: 226
512. *Diller W F* (1985) Zbl Arbeitsmed 35: 80
513. *Mowe G* (1980) Contact Dermatitis 6: 44
514. *Rothe A* (1976) Berufsdermatosen 24: 7
515. *Liden C* (1980) Contact Dermatitis 6: 301
516. *Woodbury J W* (1956) Ind Med Surg 25: 540
517. *Porter C V, Higgins R L, Scheel L D* (1975) Am Ind Hyg Ass J 36: 159
518. *O'Brien I M, Newman-Taylor A J, Burge P S, Harries M G, Fawcett I W, Pepys J* (1979) Clin Allergy 9: 7
519. *Brooks S M, McKay R T* (1981) Am Rev Respir Dis 123: 133
520. *Kondratev G G, Mustaev R K* (1969) Gig Primen Polim Mater Izdelii Nikh 1: 290
521. *Emmett E A* (1976) J Occup Med 18: 802
522. *King C M* (1980) Contact Dermatitis 6: 353
523. *Israeli R, Smirnow V, Sculsky M* (1981) Int Arch Occup Environ Health 48: 179
524. Hauptverband der Gewerblichen Berufsgenossenschaften (1981), Berufsgenossenschaftliche Grundsätze für arbeitsmedizinische Vorsorgeuntersuchungen, Isocyanate G 27, 2nd edn May 1981, Gentner, Stuttgart
525. *Fabbri L, Dal Vecchio L, Maestrelli P, Mapp C* (1984) Am Rev Respir Dis 129: 159
526. *Fabbri L, Di Giacomo R, Dal Vecchio L, Zocca E, De Marzo N, Maestrelli P, Mapp C* (1985) Bull Eur Physiopathol Respir 21: 421
527. *Fabbri L M, Chiesura-Corona P, Dal Vecchio L, Di Giacomo R, Zocca E, De Marzo P, Maestrelli P, Mapp C* (1985) Am Rev Respir Dis 132: 1010
528. *Boschetto P, Fabbri L, Zocca, Milani G, Pivirotto F, Dal Vecchio A, Plebani M, Mapp C* (1987) J Allergy Clin Immunol 80: 261
529. *Boschetto P, Zocca E, Bruchi O, Cappellazzo G, Milani G, Pivirotto F, Mapp C, Fabbri L* (1987) In: B Samuelsson et al (eds) Adv Prostaglandin, Thromboxane and Leukotriene Research, Raven, New York, vol 17 p 1080
530. *Moscato G, Gherson G, Biscaldi G, Candura F* (1983) Med Lavoro 74: 247
531. *General papers and review articles*
   Friebel H, Luechtrath H (1955) Arch Exper Path Pharmacol 227: 93
   Zapp J A (1957) Arch Ind Health 15: 324
   Peters J M, Murphy R L H (1970) Ann Intern Med 73: 654
   Peters J M (1970) Proc R Soc Med 63: 372
   Julien G et al (1970) Arch Mal Prof 31: 4
   Peters J M, Murphy R L H (1971) Am Rev Respir Dis 104: 432
   Di Bosco M M (1972) Infortuni Malattie Professionali, p 1032
   Peters J M (1974) Ann NY Acad Sci 221: 44

Nava C, Arbosti G, Briatico G, Cirla A M, Marchisio M, Zedda S (1975) La Ricerca Clin Lab 5: 135
Carroll K B, Secombe C J P, Pepys J (1976) Clin Allergy 6: 99
Karr R M, Davies R J, Butcher B T, Lehrer S B, Wilson M R, Dharmarajan V, Salvaggio J E (1978) J Allergy Clin Immunol 61: 54
Cirla A, Sala C (1978) Med di Lavoro 69: 393
Hardy H L, Devine J M (1979) Ann Occup Hyg 22: 421
Weill H (1979) J Allergy Clin Immunol 64: 662
Pepys J (1980) J Allergy Clin Immunol 66: 179
Sartorelli E (1982) Ettore Majorana Int Sci Ser Life Sci 6: 481
Pepys J (1982) J. Occup Med 24: 534
Butcher B T (1982) Eur J Respir Dis 63 suppl 123: 82
Grosclaude M (1983) Médicine et hygiène 41: 1538
Ameille J, Brochard P, De Palmas J, Proteau J (1985) Arch Mal Prof 46: 385
Butcher B T (1985) Folia Allergol Immunol Clin 32: 29
Vergnon J M et al (1985) Arch Mal Prof 46: 321
Crepet M, Mapp C E (1985) Folia Allergol Immunol Clin 32: 3
Kay S (1985) Food Chem Toxic 23: 411
Diller W F, Brochhagen F K, Slawyk W (1985) Zbl Arbeitsmed 35: 88
Diller W F (1987) Proceedings Polyurethane World Congress p 193
Goedhart F J M, Plieksier K (1987) Kunststof & Rubber 40: 21
Musk A W, Peters J M, Wegman D H (1988) Am J Ind Med 13: 331
Diller W F (1988) Bibra Bull 27: 133

532. Epidemiological studies on individuals in TDI foam plants
Gandevia B (1963) Br J Ind Med 20: 204
Gandevia B (1964) Aust Ann Med 13: 157
Peters J M, Murphy R L H, Ferris B G (1969) XVI Int Congr Occup Health (Tokyo) 11: 623
Musk A W, Peters J M, DiBerardinis, Murphy R L (1978) Am Rev Respir Dis 117: 252
Zedda S, Cirla A M, Sala C (1975) Securitas 60: 511
O'Brien I M, Harries M G, Burge P S, Pepys J (1979) Clin Allergy 9: 1
Malo J L, Zeiss C R (1982) Am Rev Respir Dis 125: 113
Holness D L, Broder I, Corey P N, Booth N, Mozzon D, Nazar M A, Guirguis S (1984) J Occup Med 26: 449
Liu J, Zhang Y, Wang S, Hu W, Qian Y (1985) Zhonghua Yufanggyixue Zazhi 19: 209
Alexandersson R, Hedenstierna G, Randma E, Rosen G, Swensen A, Tornling G (1985) Int Arch Occup Environ Health 55: 149
Miki T, Shima S, Tachikawa S, Kato Y, Yoshida T, Ito T, Taniwaki H, Hidaka K, Yuri T, Hosoda H (1985) J Sci Labour 61: 599
Musk A W, Peters J M, Bernstein L (1985) J Occup Med 27: 917
Taniwaki H, Shima S, Tachikowa S, Kato Y, Yoshida T, Ito T, Nagaoka K, Yuri T, Hosoda H, Miki T (1987) Japan J Ind Health 29: 130

533. Epidemiological studies on individuals in TDI plants
Weill H, Salvaggio J E, Neilson A, Butcher B T, Ziskind M M (1975) Envir Health Persp 11: 101
Butcher B T, Salvaggio J E, Weill H, Ziskind M M (1976) J Allergy Clin Immunol 58: 89
Knudson R J, Slatin R C, Lebowitz D, Burrows B (1976) Am Rev Respir Dis 113: 587
Butcher B T, Jones R N, O'Neil C E, Glindmeyer H W, Diem J E, Dharmarajan V, Salvaggio J E (1977) Am Rev Respir Dis 116: 411
Sharanova Z V, Penknovich A A, Dorofeeva N D, Kryzhanovskaya N A, Melnikova N D, Volkova I D, Golova I A, Arzyaeva E Y, Klimova E I, Maze E N (1982) Gig Tr Prof Zabol 1: 16
Kryzhanovskaya N A, Ermakova G A, Sharonova Z V, Volkova I D, Lyuro S D, Filatova V S (1987) Gig Tr Prof Zabol 5: 27

534. Epidemiological studies on individuals handling MDI
Konzen R B, Craft B F, Scheel L D, Gorski C H (1966) Am Ind Hyg Assoc J 27: 121
Lutier F et al (1970) Ouest-Médical 20: 823
Tanser A R, Bourke M P, Blandfort A G (1973) Thorax 28: 596
Martin F, Fichet D, Arsanian G, Leloup M C (1982) Arch Mal Prof 43: 481
Tse K S, Tabona M, Sherkin M, Kijek K, Chan H, Chan-Yeung M (1983) J Allergy Clin

Immunol 71: 158
Johnson A, Chan-Yeung M, Maclean L, Atkins E, Dyboncio A, Cheng F, Enarson D (1985) Br J Ind Med 42: 94
Low I, Mitchell C (1985) Br J Ind Med 42: 1011
Pham Q T et al (1986) Arch Mal Prof 47: 311
Filatova V S, Vinogradova V K, Belyakov A A, Mironov L A, Kalyaganov P I, Gorbunova N A, Gnelitsky G I, Volkova I D, Ermakova G A, Alieva T I, Zoloto L V (1986) Gig Tr Prof Zabol 12: 26

535. Epidemiological studies on individuals handling components for polyurethane coatings
Filatova V S, Kurando T B, Tubina A J (1968) Gig Tr Prof Zabol 12: 3
Gaffuri E, Brugnone F (1971) Med Lavoro 62: 151
Cirla A M, Aresini G, Briatico G, Inzoli S, Nava C, Zedda S (1975) Med Lavoro 66: 5
Mapp C, Moro G, Fabbri L, Crepet M (1979) Med Lavoro 70: 203
Venables K M, Dally M B, Burge P S, Pickering C A C, Newman-Taylor A J (1985) Brit J Ind Med 42: 517
Seguin P, Allard A, Cartier A, Malo J L (1987) J Occup Med 29: 340
Alexandersson R, Hedenstierna G, Plato N (1987) Arch Envir Health 42: 367

536. Miscellaneous aspects of epidemiology
Nava C, Briatico G (1980) Med Lavoro 71: 305
Brooks S M, McKay R T (1981) Am Rev Respir Dis 123: 133
Cockcroft D W (1982) Ann Allergy 48: 93
Baur X, Dewair M, Fruhmann G (1982) Eur J Respir Dis 62: 128
Paggiaro P L, Loi A M, Rossi O, Ferrante B, Pardi F, Roselli M G, Baschieri L (1984) Clin Allergy 14: 463
Holness D L, Broder I, Corey P N, Booth N, Mozzon D, Nazar M A, Guirguis S (1984) Paggiaro P L, Innocenti A, Bacci E, Rossi E, Talini D (1985) Thorax 41: 279
Baur X (1986) Allergologie 9: 487

537. Provocative inhalation challenge – techniques and experiences
Horowitz J G, Chester E H, Gerblich A A, Fleming G M, Schwartz H J (1978) J Lab Clin Med 93: 634
Gerblich A A, Horowitz J G, Chester E H, Schwartz H J, Fleming G M (1979) J Allergy Clin Immunol 64: 658
Trevisan A, Moro G (1981) Int Arch Occup Envir Health 49: 129
Barkman H W, Banks D E, Hendrick D J, Jones R N, Abdel Kader H M, Hammad Y Y, Glindmeyer H W, Weill H (1984) Chest 86: 340
Mapp C, Polato R, Maestrelli P, Fabbri L (1984) Med Lavoro 75: 115
Mapp C, Polato R, Maestrelli P, Hendrick D J, Fabbri L (1985) J Allergy Clin Immunol 75: 568
Mapp C, Dal Vecchio L, Boschetto P, Fabbri L (1985) Annals Allergy 54: 424
Mapp C, Dal Vecchio L, Boschetto P, De Marzo N, Fabbri L (1986) Eur J Respir Dis 68: 89
Mapp C, Di Giacomo R, Omini C, Broseghini C, Fabbri L (1986) Eur J Respir Dis 69: 276
Moller D F, Brooks S M, McKay R T, Cassedy K, Kopp S, Bernstein I L (1986) Chest 90: 494

538. Mechanisms of development of isocyanate hypersensitivity and asthma
Scheel L D, Killens R, Josephson A (1964) Am Ind Hyg Assoc J 25: 179
Pepys J (1974) Ann NY Acad Sci 221: 27
Steinmetz P R, Al Awquati Q, Lawton W J (1976) Am J Med Sci 271: 40
Sharonova Z V, Kryzhanovskaya N A (1976) Gig Tr Prof Zabol 11: 27
Zedda S, Cirla A, Aresini G, Sala C (1976) Respiration 33: 14
Braman S S, Teplitz C (1978) Primary Care 5: 425
Siracusa A, Curradi F, Abbriti G (1978) Clin Allergy 8: 195
Cockcroft D W, Mink J T (1979) Can Med Assoc J 121: 602
Burge P S, O'Brien I M, Harries M G (1979) Thorax 34: 317
Burge P S, Sherwood P, O'Brien IM, Harries M G (1979) 000
Harries M G, Burge P S, Samson M, Newman-Taylor A, Pepys J (1979) Thorax 34: 762
Salvaggio J E (1979) J Allergy Clin Immunol 64: 646
Butcher B T, O'Neil C E, Reed M A, Salvaggio J E (1980) J Allergy Clin Immunol 66: 213
Baur X, Dorsch W, Fruhmann G, Roemmelt H, Roth P, Diller W (1980) Zts Arbeitsmed Arbeitssch Prophylaxe Ergon 30: 104
Baur X, Fruhmann G (1981) Chest 80: 73

Patterson R, Suszko I M, Zeiss C R, Pruzansky J J (1981) J Clin Immunol 1: 181
Butcher B T (1982) Eur J Respir Dis 123: 78
Baur X, Albrecht J, Huber R M, Kessel R, Koenig G, Roemmelt H, Fruhmann G (1982) Ber
    Jahrestag Deut Med Ges 22: 597
Burge P S, Sherwood P (1982) Eur J Respir Dis 63: 91
Butcher B T, O'Neil C E, Reed M A, Salvaggio J E (1983) Clin Allergy 13: 31
Dewair M, Baur X, Fruhmann G (1983) J Occup Med 25: 279
*Paggiaro P L, Filieri M, Loi A M, Roselli M G, Canalupi R, Parlanti A, Toma G, Baschieri
    L* (1983) Clin Allergy 13: 75
Baur X (1983) J Allergy Clin Immunol 72: 197
Patterson R, Harris K E, Zeiss C R (1983) J Allergy Clin Immunol 72: 676
Valentino M, Ruschioni A, Rocco M, Comai M, Governa M (1986) Boll Soc It Biol Sper 62: 1245
*Fabbri L, Boschetto P, Zocca E, Milani G, Pivirotto F, Plebani M, Burlina A, Licata B, Mapp
    C* (1987) Am Rev Respir Dis 136: 36
Patterson R, Hargreave F E, Grammer L C, Harris K E, Dolovich J (1987) Int Arch Allergy Appl
    Immunol 84: 93

539. *Case histories*
Swensson A, Holmquist C E, Lundgren K D (1955) Br J Ind Med 12: 50
Woodbury J W (1956) Ind Med Surg 25: 540
Jennings G H, Gower N D (1963) Lancet I: 406 (88);
Maxon F C (1964) Arch Environ Health 8: 755
Sweet L C (1968) Univ Mich Med Gr J 34: 27
Schmidt-Nowara W W, Murphy R L H, Atkinson J D (1973) Chest 63: 1039
Belin L, Hjortsberg U, Wass U (1981) Scand J Work Envir Health 7: 310
Pacheco Y, Vergnon J M, Grosclaude M, Biot N, Perrin-Fayolle M (1982) Rev Fr Allergol 22: 15
Mortillaro P T, Schiavon M (1982) Med Lavoro 73: 207
Innocenti A, Paggiaro P L (1983) Med Lavoro 74: 391
Malo J L, Ouimet G, Cartier A, Levitz D, Zeiss C R (1983) J Allergy Clin Immunol 72: 413
Fink J N (1984) J Allergy Clin Immunol 74: 1
Laitinen J, Muittari A, Sovijarvi A (1984) Duodecim 100: 220
Hargreave F E, Ramsdale E H, Pugsley S O (1984) Am Rev Respir Dis 130: 513
Baur X, Dewair M, Roemmelt H (1984) J Occup Med 26: 285
Sandridge R L, Baur X, Dewair M, Roemmelt H (1985) J Occup Med 27: 786
Beysens T, Meyer P D, Schaller M, Pauli G, Stoeckel C (1985) Rev Fr Allergol 25: 199
Innocenti A, Mariano A, Valiani M (1986) Med Lavoro 77: 191
*Banks D E, Barkman H W, Butcher B T, Hammad Y Y, Rando R J, Glindmeyer H W, Jones R N,
    Weill H* (1986) Chest 89: 389

Nitro Derivatives of Polycyclic Aromatic Hydrocarbons (NO$_2$-PAH)

Heidelore Fiedler[1] and Wolfgang Mücke[2]

[1] Lehrstuhl für Ökologische Chemie und Geochemie, Universität Bayreuth, Postfach 10 12 51, D-8580 Bayreuth
[2] Bayerisches Staatsministerium für Landesentwicklung und Umweltfragen, Rosenkavalierplatz 2, D-8000 München 81, and Ludwig-Maximilians-Universität München

Summary

Like the PAH, Nitro-PAH are universally distributed and predominantly emitted through anthropogenic activities such as combustion processes: Exhaust gases from motor vehicles are regarded as being the main source. Although present in smaller quantities than the corresponding PAH, they deserve special attention because of their biological activity: a number of them have proved to be

The Handbook of Environmental Chemistry,
Volume 3 Part G, Ed. O. Hutzinger
© Springer-Verlag Berlin Heidelberg 1991

mutagenic and carcinogenic in animal testing. Nitro-PAH are not only emitted directly but can also be formed indirectly through reactions of PAH with gaseous air pollutants.

Nitration of PAH is a very complex reaction in which the formation rate of the nitro derivatives is influenced by the medium and the presence of NO_2, H^+, hydroxyl radical, HNO_3, HNO_2, and metal ions with a high oxidation potential. Additional parameters are the surface characteristics and grain size of the particles on which the PAH are adsorbed and where they take part in further reactions.

The nitro-PAH concentrations in particles from diesel and gasoline exhaust gases lie in the same order of magnitude whereas the emission rate from diesel engines is higher, however, because they expel more particles per unit distance traveled than gasoline engines. Vehicles equipped with a catalytic converter produce substantially lower amounts of nitro-PAH. Dinitropyrene, a mutagenic, particle-bound component in emissions from incomplete combustion processes, causes between 20 and 40% of the mutagenicity. Nitro-PAH are direct mutagens in the Ames-test. Dinitro-PAH, in general, appear to be stronger mutagens than mononitro-PAH.

The most important transformation of nitro-PAH is photodecomposition to the respective arylnitrite followed by the formation of the phenoxyl radical with the elimination of dinitrogen monoxide. It further reacts to nitrohydroxy derivatives and quinones. The arylnitrite can also hydrolyze to hydroxy derivatives and nitric acid.

Sampling techniques for automobile exhaust have been developed and analytical methods for clean-up, separation, and detection of nitro-PAH are given.

1 Introduction

The formation, distribution, and effect of polycyclic aromatic hydrocarbons (PAH) have been the subject of intensive research for a number of years. A short time ago their nitro derivatives (nitro-PAH) entered into the discussion. They, like PAH are universally distributed pollutants and they predominantly emanate from anthropogenic sources [1, 2]. Their main source appears to be combustion processes and they are distributed by air masses. Although, they occur at much lower concentrations than PAH, they deserve special attention because of their biological activity: a number of them have proved to be mutagenic [3] and carcinogenic in animal testing [4]. This finding has led to concern with their concentrations in environmental compartments and also with processes of formation and transformation in the atmosphere.

Exhaust gases from motor vehicles are regarded as being the main source of these pollutants from combustion sources; evidence from Otto- and Diesel engine emissions has been obtained. The conditions for formation will be closely examined as well as the further fate of nitro-PAH in the environment.

An important question is how Diesel and Otto engine emissions are to be judged with regard to nitro-PAH (see [5]). Diesel motor emissions are classified in the German MAK-value-list[1] Section III as A2[2]. Pyrene nitro derivatives (mono-, di-, tri-, tetra-isomers) are classified as B in Section III[3] [6]). Non-carcinogenic species of nitro-PAH can contribute to the carcinogenity of other PAH by the enzyme-induction route [7].

---

[1] MAK = maximum workplace concentration.
[2] Substances which up to now in animal experiments and according to the commission have clearly shown themselves to be carcinogenic under conditions which are comparable to those experienced by people at work or conditions from which a comparison can be derived.
[3] Substances reasonably suspected of having carcinogenic potential, which need urgent clarification.

2 Theoretical Basis

2.1 Chemistry of Nitro-PAH

Nitro-PAH are the nitrated derivatives of polycyclic aromatic hydrocarbons. Their condensed ring systems come from the incomplete combustion or pyrolysis of organic materials such as fossil fuels and wood. So, for example, retene (1-methyl-7-isopropylphenanthrene) [8] from coniferous wood serves as the archetypal substance for PAH in wood smoke [9]. Nitro-PAH have been identified in environmental samples, in gasoline and Diesel exhaust gases, fly ash [10], printer's ink, urban air [11, 12], and in cigarette smoke [13]. A number of them are suspected of being carcinogenic and/or mutagenic.

The structural formulas of nitro-PAH commonly found in environmental samples are shown in Fig. 1 [14] and physical data are given in Table 1. Mononitro- and dinitro-PAH are found especially in Diesel exhaust gases, in atmospheric aerosol, and in soot [15]. They are, however, not in the gaseous form in the air but are found adsorbed onto particles of small grain size [16]. There is insufficient information on accumulation in the soil and sediments as well as for possible accumulation in tissues.

Table 1. Physical data of nitro-PAH

Substance	Molecular weight (g/mol)	Melting point (°C)	Boiling point (°C)	CAS #
1-Nitronaphthalene	173.17	54–56	304	86–57–7
2-Nitronaphthalene	173.17	74–76		581–89–5
2-Nitrofluorene	211.22	154–157		607–57–8
4-Nitrobiphenyl	199.21	112–114		92–93–3
9-Nitroanthracene	223.23	141–143		602–60–8
1-Nitropyrene	247.06	151–152	610	5522–43–0
3-Nitrofluoranthene	247.06	156–158	570	
6-Nitrochrysene	273.08	206.5–208.5		7496–02–8
6-Nitrobenzo[a]pyrene	297.08	250–255		
3-Nitroperylene	297.08	209–209.5		
1,8-Dinitronaphthalene	218.17	171–172		602–38–0

Structures of nitro-PAH commonly found in environmental samples are shown in Fig. 1 [14].

2.2 Mechanisms of Formation

The basic PAH skeleton of nitro-PAH is known to result from incomplete combustion or pyrolysis of organic substances containing carbon and hydrogen. When the supply of oxygen to the combustion of organic substances, e.g. fossil fuels, is adequate, burning to the end products of carbon dioxide and water is

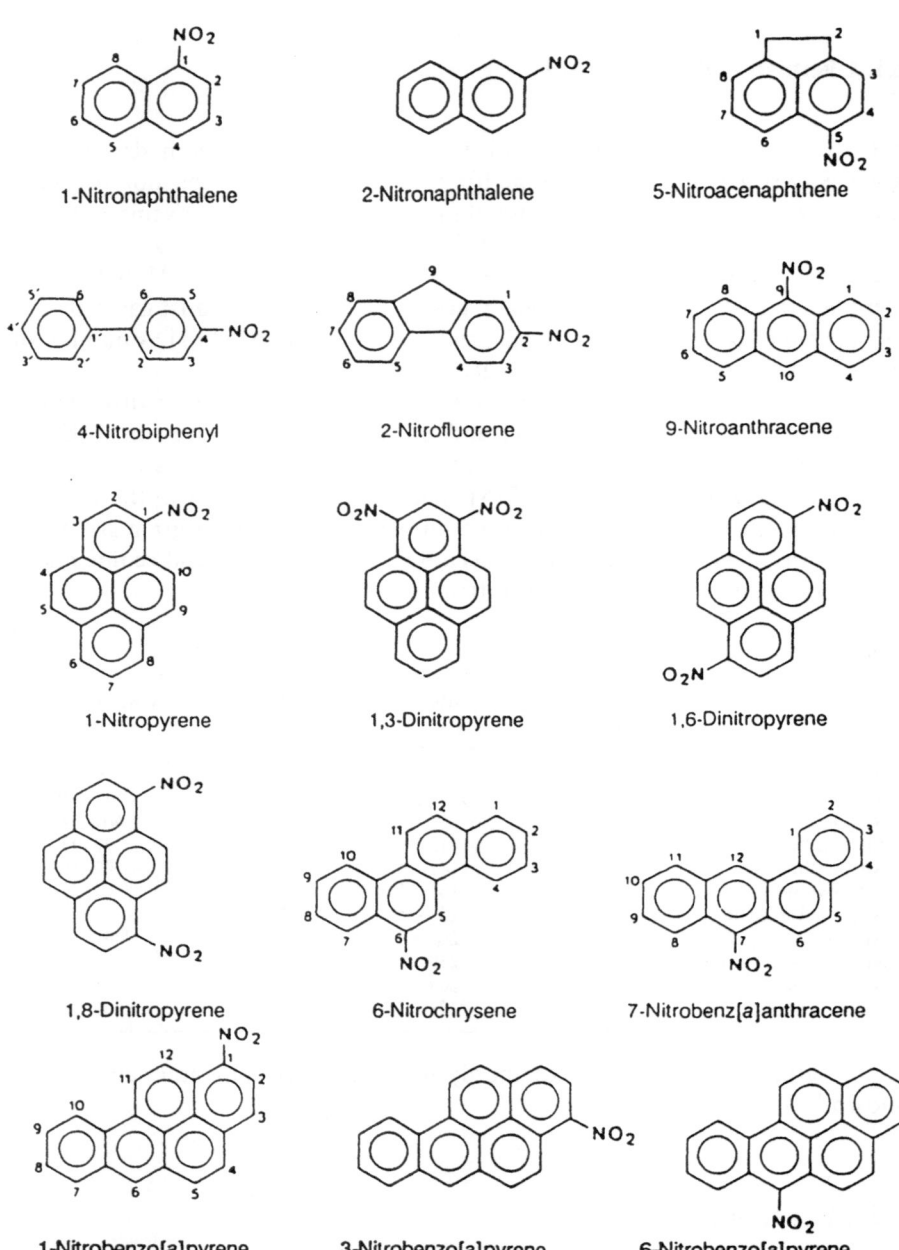

Fig. 1. Structures of nitro-PAH commonly found in the environment [14]

Table 2. Typical concentrations of NO_x, N_2O, HNO_3, O_3 and gaseous non-methane hydrocarbons in the atmosphere [20]

Substance	Typical concentrations in the air	
	Urban	Background
NO_x	40[a] ppb	0.1–2 ppb
N_2O	330 ppb	33 ppb
HNO_3	0.5 ppb	0.01 ppb
O_3	80[a] ppb	20–40 ppb
Non-methane hydrocarbons	0.1–2 ppm	0.1 ppm

[a] The concentrations can reach 260 ppb O_3 and 150 ppb NO_x in special circumstances of immission

more complete and the proportion of PAH produced is smaller. Incomplete combustion or pyrolysis is due to an insufficient supply of oxygen. In general, gases can be mixed homogeneously so that there are no micro-regions with a lack of oxygen. For this reason there are always typical profiles at particular temperatures independent of the type of fuel. Generally the PAH pattern mainly depends on the temperature of burning [17]. As will be shown later, there are various possibilities of nitration and the formation of nitro-PAH. It has been known for several years that nitro-PAH are not only emitted directly but also formed by reactions of PAH with gaseous air pollutants [18, 19, 23]. Table 2 shows typical concentrations of oxidizing gases and of non-methane hydrocarbons.

PAH can be nitrated relatively easily. Under mild reaction conditions, substitution occurs on the most reactive carbon atoms, whereby the kinetically prefered isomer is formed. In addition to the main product, other isomers or more highly nitrated PAH derivatives are always formed [14]. The reaction mechanisms which lead to the formation of nitro-PAH have not yet been definitively established. It is also not completely clear whether nitro-PAH are formed during combustion (e.g. in the Diesel engine), in the gas phase or only after adsorption onto particles. Since PAH are products of incomplete combustion of organic materials, nitro-PAH can, in the presence of nitrogen, also be produced. Nitro-PAH could also be formed in the air from PAH and nitrogen oxides. This theory is supported by the fact that the commonly occurring nitro-PAH are derivatives of PAH that are only detected in low concentrations in the surrounding air. Furthermore, it is known that nitro-PAH are formed as an artifact during sampling. This makes the evaluation of available experimental findings extraordinarily difficult. The reactions in the gas phase must, however, be distinguished from those in the particle phase because the latter can be subject to many heterogeneous and catalytical effects on the particle surface.

2.2.1 Reactions with Hydroxyl Radicals

The transformation of PAH to their nitro derivatives proceeds in the gas phase by radical processes. The radical mechanism has been demonstrated in experiments in the smog chamber and also in an atmosphere with high contents of photochemical oxidants and nitrogen oxides [1]. This reaction is shown with toluene as an example in Fig. 2, whereby, after the addition of a hydroxyl radical to the aromatic ring system, nitrogen oxide is first added and then water is eliminated to give the nitro derivative of the arene. The possible formation of isomers is not shown in Fig. 2.

At high temperatures another mechanism is conceivable by which the transformation of the aromatic hydrocarbon in a phenolic radical proceeds via hydroxyl radicals and nitrogen dioxide. Such a mechanism for the formulation of mononitro-PAH in stack gases and automobile exhausts is possible but has not yet been proven.

The average concentration of hydroxyl radicals in the atmosphere at 60 °N latitude is about 1×10^6 molecules/cm^3 in the summer and much less than 0.5×10^5 molecules/cm^3 in the winter months [21]. Considering only the reactions with hydroxyl radicals the half-life of benzene will be about 6 days in summer and much longer than 8 months in winter [2]. The reactivities of PAH towards hydroxyl radicals should be expected to be higher than that of benzene but not by more than two orders of magnitude [22]. Although a high concentration of hydroxyl radicals can be observed in exhaust gases, the hydroxy-cyclohexadienyl intermediate product shown in Fig. 2 is either not formed at temperatures higher than 100–150 °C or its disproportionation takes place [18, 23, 24]. The more probable reaction mechanism at higher temperatures appears to be the one shown in Fig. 3.

It is known that 2-nitropyrene and 2-nitrofluoranthene are not emitted directly from the various sources but rather formed through transformation processes [25]. Gas phase investigations have shown that 2-nitropyrene and 2-nitrofluoranthene (with small quantities of 7- and 8-nitrofluoranthene) are formed from pyrene and fluoranthene in the presence of hydroxyl radicals and NO$_x$ [26]. The radical mechanism shown in Fig. 4 demonstrates how a hydroxyl radical is added at C^1 of the pyrene and then nitrated at C^2 leading to the formation of 2-nitropyrene.

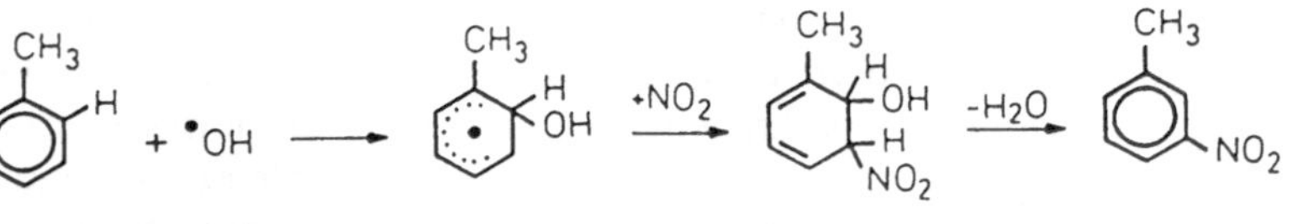

Fig. 2. Mechanism of nitroarene formation using toluene as an example [1]

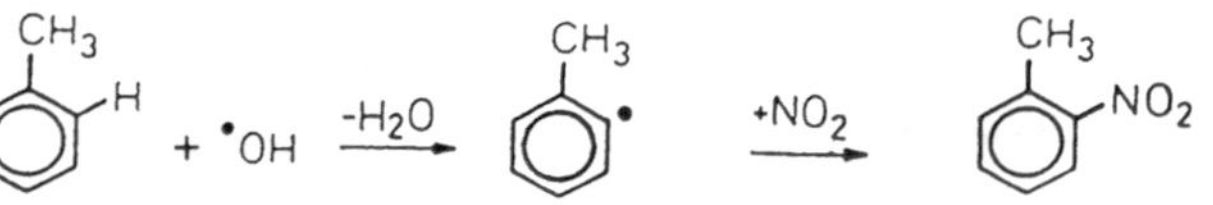

Fig. 3. Mechanism of nitroarene formation in the gas phase at higher temperatures using toluene as an example [1]

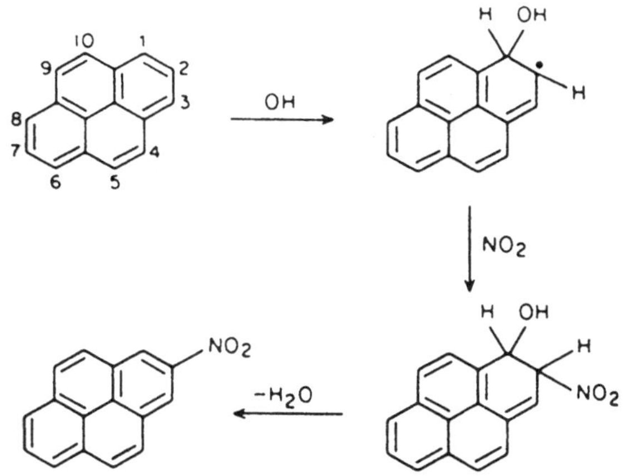

Fig. 4. Mechanism of 2-nitropyrene formation from nitropyrene in the gas phase [26]

A similar mechanism of formation is assumed for 2-nitrofluoranthene, where the initial attack of the hydroxyl radical at the C^1 of fluoranthene as well as the attack at the C^3 (the most reactive in the molecule) can occur. In a second step nitration will give 2-nitrofluoranthene. A radical attack at the C^2 of the fluoranthene is expected due to the low reactivity of C^2 – and this is verified by the fact that neither 1- nor 3-nitrofluoranthene has been detected in the presence of NO_x.

Nitrophenol derivatives of PAH can be formed as shown in Fig. 5. The half-life of the formation process of nitrophenol via the phenoxyl radical is 0.07 seconds [27]:

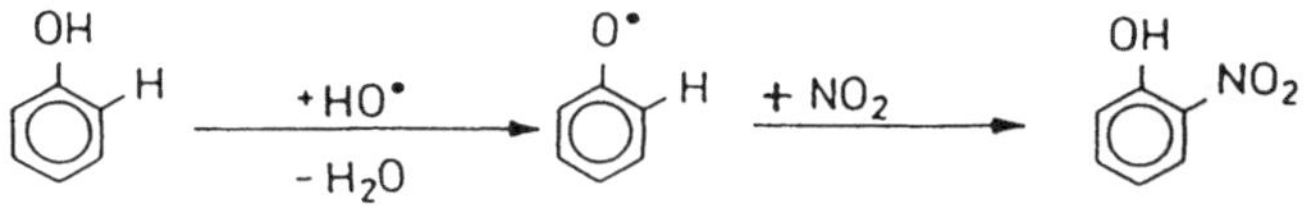

Fig. 5. Formation mechanism of nitrohydroxy aromatic compounds [27, 28]

2.2.2. The Reaction with Nitrogen Trioxide

The reaction of PAH with hydroxyl radicals is not the only important pathway for the formation of phenoxyl radicals. At night-time at low relative humidity (less than 60%) more than 100 ppt of nitrogen trioxide has been observed in photochemically polluted air [29, 30]. Under these conditions the nitration of PAH with nitrogen trioxide may be a significant mechanism for the formation of nitro-PAH. Generally NO_2-PAH formation takes place preferentially in summer and at night. During darkness with low humidity NO_2 reacts with ozone according to Eq. 1. In cases of high humidity, the dimer N_2O_4 is formed from the monomer NO_2 and in turn this is hydrolysed to form nitrous and nitric acids. Nitration can also occur directly with nitric acid.

However, by the nitration of PAH, it is not the NO_3 which reacts but rather N_2O_5 formed from NO_3 and NO_2 (Eq. 2). This is very unstable and reacts immediately with any PAH molecules present. With an occurrence of approximately 10 μg/kg about 5% of the existing pyrene was nitrated [31].

$$NO_2 + O_3 \longrightarrow NO_3 + O_2 \tag{1}$$

$$NO_3 + NO_2 \longrightarrow N_2O_5 \tag{2}$$

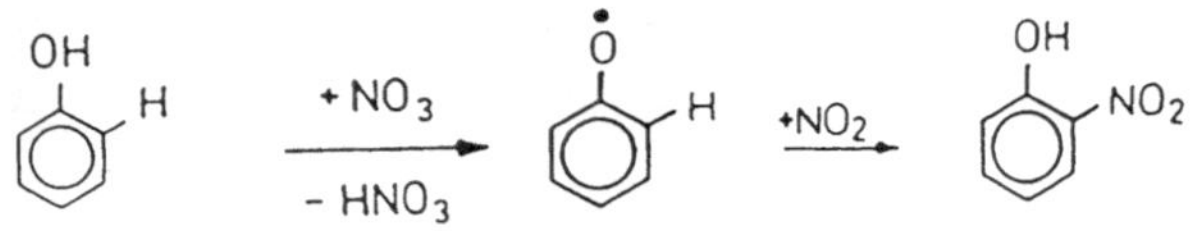

Fig. 6. Formation mechanism of nitrohydroxy aromatic compounds by NO_3 using phenol as an example (simplified model) [27]

2.2.3 Reactions of Particle-Bound PAH

In contrast to the gas phase, where the radical reactions are strongly influenced by the wavelength and the intensity of the light, the formation of nitro derivatives from PAH adsorbed on particles is not affected by the light conditions [32]. Darkness, diffused daylight, or light from a mercury vapour lamp have no appreciable effect on the reaction rates or the composition of the products.

For particle-bound PAH catalytic effects are much more important due to the heterogeneous composition and surface activity of the particles. The decisive factor for the formation rate of the nitro derivatives is the constitution of the carrier, e.g. soot, fly ash, silica, aluminium oxide. It has been demonstrated that each carrier material on which the PAH is bound acts as a sorbent and at normal temperatures adsorbs traces of NO_2 (up to the ppm region) and releases them again at higher temperatures. The formation rates of 1-nitropyrene and 6-nitrobenzo[a]pyrene for example are very low when pyrene and benzo[a]pyrene are adsorbed on carbon-containing particles from the exhaust gases from a 4-cylinder engine. In contrast, the formation rates are high when the above mentioned chemicals are adsorbed on silica gel, aluminium oxide (neutral), or fly ash [32, 33]. Correspondingly high transformation rates to form the nitro derivatives are found when benzo[a]pyrene and perylene are adsorbed onto glass fibers [34] whereas the transformation of PAH adsorbed on soot from the incomplete combustion of ethylene is relatively slow [35]. It appears that the formation rate becomes faster with increasing polarity and/or decreasing basicity of the adsorbent.

The chemical structure of the adsorption sites is not yet clearly understood. Some results point to the fact that PAH bind especially to particles with a high carbon content [35, 36]. A strong binding of PAH on particles is indicated by the high extraction temperatures and long extraction times[1] which are necessary for quantitative extraction of PAH from complex matrices such as diesel emissions or printer's ink. With diluted exhaust gases, it can be shown that the distribution of PAH between adsorption on particles or the gas phase is not influenced by carbon-containing particles alone but also by raised concentrations of lead halides and oxides [37].

Although the nitro-PAH formation rate on newly formed carbon-containing particles is low this can be increased by aging processes of the particles. The aging of a particle is caused mainly by oxidation processes in the atmosphere. Thereby additional hydrophilic centres and free acids on the particle surface are formed and these catalyze the reaction of the PAH with nitrogen oxides [38, 39].

According to the classic mechanism of electrophillic aromatic nitration, the nitroarene is formed via the nitroarenium ions [40]. However, for this mechanism only small variations in reactivity would be expected for the various PAH and in practice this is not the case. The differences in reactivity which occur are an indication that the transformation of PAH to their nitro derivatives is via radical cations and carbenium ions. The low concentration of nitro-PAH in the atmosphere also points to the formation mechanism having a radical cation as an intermediate stage. These intermediate stages were demonstrated experimentally [41, 42]. The transformation of PAH via this intermediate stage of the radical cation can be formulated as shown in Fig. 7.

In a comparative study [43] the behaviour of three particle-bound PAH towards gaseous pollutants has been investigated. Benzo[a]pyrene has been

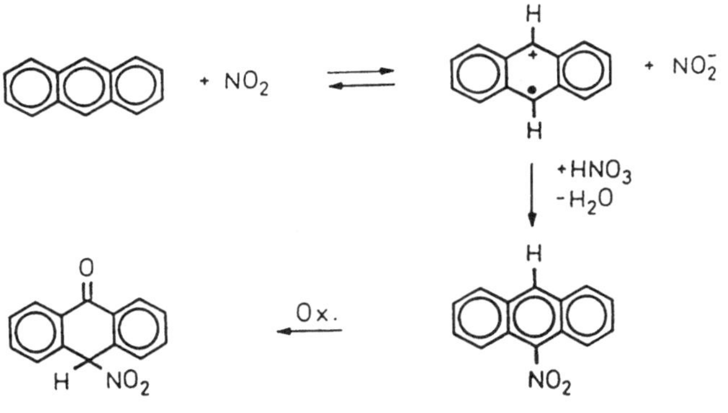

Fig. 7. Transformation of particle-bound PAH to their nitro derivatives via radical cations [41, 42]

---

[1] According to the VDI (German Association of Engineers) Regulation 3872 the extraction time is only 30 min (with backflow).

chosen as a reference and as the best characterized substance; perylene, an isomer of BaP was included to investigate isomer specific behaviour; and 1-nitropyrene was added because it is held responsible for the mutagenic effect of Diesel exhaust gases. In smog chamber experiments, when 110 ppb NO_2 were introduced, no reduction in the concentration of PAH could be detected.

The nitrogen species involved in the nitration of PAH are mainly NO_3^-, HNO_2, HNO_3, NO_2, N_2O_4, N_2O_5 [31], alkyl nitrites, and alkyl nitrates [2]. NO, in contrast, is not a reagent in the transformation of PAH to nitro-PAH [1].

Basically, the low concentrations of nitro-PAH in the atmosphere (10–100 times lower than PAH) [44] are in conflict with the easily accomplished laboratory transformation of PAH to the nitro derivatives. Until now, it has not been possible to describe the chemical and physical factors responsible for the differences in reactivity. For PAH which are already substituted it is true that:

— Substituents which act as electron donors accelerate the transformation rate.
— Weak electron-accepting substituents such as Cl cause a delay of one order of magnitude, strong electron-accepting substituents such as NO_2 slow the transformation by two orders of magnitude [2].

Experiments are necessary to further elucidate the reaction mechanisms in the atmosphere and in automobile engines under experimental conditions which are as realistic as possible. For automobile engines, dynamic and catalytic effects during the combustion must be taken into consideration.

To summarize, it can be stated that the transformation from PAH to nitro-PAH in the atmosphere takes place only to a small extent [44], i.e. the greater proportion is formed during combustion. The nitration of PAH is a very complex reaction, in which the formation rate of the nitro derivatives is influenced by the medium and the presence of NO_2, H^+, hydroxyl radicals, HNO_3, HNO_2, and metal ions with a high oxidation potential [44, 45]. Additional parameters are the surface characteristics and grain size of the particles on which the PAH are adsorbed and where they take part in further reactions.

Of practical significance is the artificial formation of nitro-PAH from PAH in experimental investigations. It has been shown with the collecting filter that increasing amounts of nitro-PAH are formed with longer collection periods. This has been demonstrated in detail using 1-nitropyrene as the example [46, 47].

The values shown in Fig. 8 were obtained by the procedure of the test cycle US-72-test (warm start; 4–6 measurements). It can be seen that for collection times under 46 minutes, corresponding to two US-72-tests, less than 20% of the total emission (on the soot particles) can be traced back to an artifact formation. The major part of the nitropyrene has already been formed in the engine or the exhaust system. With extending sampling periods noticeable changes in the relative concentrations will occur due to reactions on the filter. This creates problems in obtaining large amounts of soot, e.g. for biological tests. This

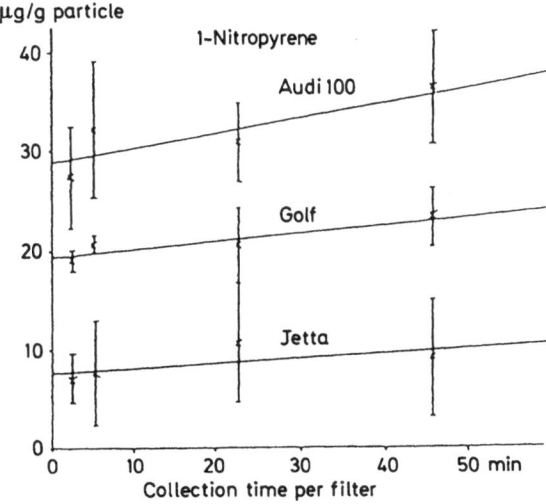

Fig. 8. Concentration of 1-nitropyrene in the particle emission from three Diesel automobiles (in µg/g particle) in relation to the collection period (by linear extrapolation to "zero" minutes of collection; the so-called baseline emission, i.e. the emission rate with practically no artifacts, is determined) [47]

problem is intensified under operating conditions with high temperature exhaust gases (e.g. with a constant driving speed of 80 km/h) and for collecting methods with strong oxidizing conditions, e.g. with electrostatic separation of the particles which leads to an increase in NO_2-formation due to ozone [47].

3 Sources of Nitro-PAH

The main sources for mononitro-PAH in the atmosphere are [2, 44]:

1. Emissions from industrial combustion processes, power stations, gasoline and Diesel engines, and also forest fires [22].
2. Ionic nitration of PAH by NO_2 and HNO_3 in the atmosphere.
3. Radical induced nitration of PAH.

The emission of nitro-PAH from stationary combustion sources is very significant because they are mainly responsible for the expulsion of airborne particles. In the USA, approximately 8×10^6 tonnes of airborne particles come from these sources. This is 90% of the total mass [48]. Model calculations in the USA, based on a prognosis of an increase in the number of Diesel vehicles, predict that the contribution of Diesel emissions to the future total emissions of nitro-PAH is very low, namely only 0.0001 $\mu g/m^3$ [18, 23]. To what degree the Diesel-engine-produced PAH, which can further react in the atmosphere to their corresponding nitro derivatives, were taken into consideration is not shown in this study. This needs thinking about, especially with the increasing atmospheric concentrations of HNO_3 which causes the transformation of PAH to their nitro derivatives.

4 Occurrence in the Environment

There are extensive data on the occurrence of PAH in the atmosphere: more than 100 PAH have been detected. Most measurements are concerned with the concentration of benzo[a]pyrene (BaP), although, because of differing methods of collection and analysis, results (especially from the 1960s) should only be conditionally compared with each other. However, one can conclude that in winter the BaP concentration is higher by a factor of 10 than in summer, higher in cities than in rural areas, and in many areas of Europe has considerably diminished during the last 20 years. The results of measurements have shown that even at places with high traffic density the contribution of BaP from automobile exhaust is only in the range of a few ng/m^3 BaP compared to the total amount of BaP released [49]. Measurements taken in Bayreuth gave a concentration of 0.3 ng/m^3 [50]. By far the greatest contribution to pollution by benzo[a]pyrene and other PAH in the FRG is coal-fired stoves in the home (56.3%) and also coke-producing plants (30.8%). Motor vehicles contribute 13%, but the contributions from coal-fired power stations (0.006%) and oil-fired heating (0.1%) are almost insignificant in comparison [51].

PAH bound to fine particles may cause pollution in densely populated areas: the average pollution by dust as a whole in Berlin (West) is between 100 and 150 $\mu g/m^3$ and for benzo[a]pyrene (BaP), which is often taken as a guide to the whole group of PAH, between 7 and 12 ng/m^3, The BaP pollution then exceeds the local "orientation value" from the German Environmental Agency which is 10 ng/m^3 [52]. With the aid of calculations using the normal balance of the chemical elements one can use the emission and immission measurements to estimate quantitively the immission contributions from various sources. The contribution from motor vehicles to the total dust and to the PAH immissions is at locations near to traffic double that of locations remote from traffic. The mean values in Berlin are: for total dust 28 $\mu g/m^3$, for benzo[a]pyrene 1.6 ng/m^3, for benzo[e]pyrene 1.2 ng/m^3, for benzo[ghi]perylene 3.1 ng/m^3, for benzo[k]fluoranthene 0.6 ng/m^3, for perylene 0.3 ng/m^3, and for anthracene 0.5 ng/m^3. In Berlin, about 20% of the dust immission and between 15 and 25% of the investigated PAH-immission is caused by traffic, whereas in Cologne and Düsseldorf approximately half of the dust immission and practically all of the PAH-immission is caused by emissions from motor vehicles [52].

The concentrations of nitro-PAH are always much lower than the corresponding PAH. Rough estimates of the concentrations of mononitro-PAH in the atmosphere have values 10–100 times lower than the more commonly occurring carcinogenic PAH [2]. The ratio of atmospheric, particle-bound 1,6-dinitropyrene to benzo[a]pyrene can vary from 1:5000 to 1:15000 [53]. In one investigation the concentrations of 1-nitropyrene and 2-nitrofluoranthene in electrically heated living rooms of non-smokers 0.002–0.004 ng/m^3 and 0.005–0.008 ng/m^3 [28]. In living rooms belonging to smokers and heated with coal stoves the concentrations were 0.005–0.56 ng/m^3 and 0.028–0.167 ng/m^3. In colour toners for photocopiers up to 10 μg/kg of 1,3-; 1,6-; 1,8-dinitropyrene were found [54]. In waste water from gas stations, 1-nitropyrene has been identified [14, 29, 55].

Table 3. Nitro-PAH detected in environmental samples (Compilation in: [56, 57])

Substance	Source
1-Nitropyrene	Airborne: Santiago/Chile, California/USA
	Fly ash
	Waste water/Gasoline station
	River sediment
	Diesel emission
	Kerosene heater, fuel gas and liquified gas burner emissions
	Air particulate/Upper Franconia (W. Germany)
	"Yakitori" grilled chicken
2-Nitropyrene	Ambient air
1,3-Dinitropyrene	Kerosene heater, fuel gas and liquified gas burner emissions
1,6-Dinitropyrene	Airborne: Santiago (Chile)
	Diesel emissions
	Kerosene heater, fuel gas and liquified gas burner emissions
1,8-Dinitropyrene	Airborne/Santiago (Chile)
	Diesel emissions
	Kerosene heater, fuel gas and liquified gas burner emissions
1-Nitro-3-acetoxypyrene	Diesel emissions
1-Nitro-6-acetoxypyrene	Diesel emissions
1-Nitro-8-acetoxypyrene	Diesel emissions
1-Nitro-3-hydroxypyrene	Diesel emissions
1-Nitro-6-hydroxypyrene	Diesel emissions
1-Nitro-8-hydroxypyrene	Diesel emissions
6-Nitrobenzo[a]pyrene	Diesel emissions
2-Nitrofluorene	River sediment
	Kerosene heater, fuel gas and liquified gas burner emissions
2-Nitro-9-fluorene	Diesel emissions
3-Nitrofluoranthene	Diesel emissions
	Air particulate/Upper Franconia (W. Germany)
2-Nitrofluoranthene	Ambient air
1,5-Dinitronaphthalene	Kerosene heater, fuel gas and liquified gas burner emissions
1,8-Dinitronaphthalene	Kerosene heater, fuel gas and liquified gas burner emissions
6-Nitrochrysene	Air particulate/Upper Franconia (W. Germany)
9-Nitroanthracene	Diesel emissions

In a rural area 30 km west of Copenhagen the following concentrations of nitro-PAH were found in the air:

Table 4. Mean values from 30 24-h samples of nitro-PAH west of Copenhagen [58]

9-Nitroanthracene	0.03 ± 0.01 ng/m^3
1-Nitropyrene	0.009 ± 0.005 ng/m^3
10-Nitrobenz[a]anthracene	0.014 ± 0.007 ng/m^3

The following tables give the atmospheric concentrations of particle-bound PAH and nitro-PAH measured in Kawasaki and Tokyo (Japan) and Beijing (China).

Table 5. Atmospheric concentrations of PAH (ng/m^3) in Japan and China (month/year of investigation) [59]

Substance	Tokyo[1] 12/84	8/84	Beijing[2] 2/83	7/83	Beijing[3] 2/83	7/83
Fluoranthene	5.5	0.1	60	2.8	12	0.7
Benzofluoranthenes	12	0.4	45	1.3	10	2.3
Benzo[a]pyrene	5.5	0.2	35	1.7	4.0	1.1
Indeno[1,2,3-cd]pyrene	22	1.5	51	8.3	10	3.8
Benzo[ghi]perylene	11	1.4	47	6.9	7.6	3.0
Anthanthrene	4.0	< 0.1	13	< 0.3	0.8	0.3
Dibenzo[ae]pyrene	7.9	0.4	24	3.1	3.4	1.7

[1] Suburb; [2] Urban district (roadside, heavy goods traffic); [3] Rural area (clean air region)

Table 6. Atmospheric concentrations of nitro-PAH (ng/m^3) in Japan and China (month/year of investigation) [59]

Substance	Kawasaki[1] 10/83	Tokyo[2] 12/84	8/83	Beijing[3] 2/83	7/83	Beijing[4] 2/83	7/83
Nitrobiphenyl	25	22	15	76	140	140	110
Nitroacenaphthene	< 1	5	2	6	75	< 2	6
2-Nitrofluorene	71	24	50	190	290	36	79
9-Nitroanthracene	890	750	120	840	650	500	530
Nitrofluoranthene (isomer)	390	310	31	260	650	250	95
1-Nitropyrene	320	110	< 7	500	74	390	45

[1] Industrial area (roadside); [2] Suburb; [3] Urban district (roadside, heavy goods traffic); [4] Rural area

A comparison of the concentrations of dinitropyrenes in the atmosphere with those in Diesel emissions in the following table shows that the Diesel exhaust gases contain 1,000 times the amount in the air.

Table 7. Comparison of dinitro-PAH concentrations in the exhaust gas of Diesel vehicles and in the atmosphere [59]

	Veh 1[3]	Diesel (ng/m^3) Veh 2[4]	10/83	Atmosphere[2] (pg/m^3) 11/84	Sum/81[5]	Atmosphere[1] (pg/m^3) Win/82[6]	Sum/82[7]
1,3-DNP	1.0	5.3	2.3	1.4	—	—	—
1,6-DNP	2.2	3.2	5.3	1.2	0.3–2.2	< 0.1	1.28–9.26
1,8-DNP	2.9	3.3	11	2.6	0.22–0.39	< 0.2	0.70–2.49

[1] Results from Gibson and Tironi, 1985 [60]; [2] Kawasaki, industrial area (roadside); [3] 1,400 rpm; Vehicle load 25%; [4] 1,000 rpm; Vehicle load 75%; [5] Detroit, urban district; [6] Warren, suburb; [7] River Roudge, industrial area

The proportion of PAH with two or three rings in the exhaust condensate of Otto engines is 9.9% by mass and that with four or more rings about 3% [51]. From Diesel exhaust extracts, more than 90 nitro-PAH have been identified [61–65]. Table 8 gives an overview of the concentrations of 1-nitropyrene and 6-nitrobenzo[a]pyrene in the air and exhaust gases from motor vehicles. In addition their mutagenic activity is shown.

As can be seen in Table 8, the nitro-PAH concentrations in particles from Diesel and gasoline exhaust gases lie in the same order of magnitude. The emission rate from Diesel engines is higher, however, because they expel more particles per unit distance traveled as gasoline engines [18, 23]. From the table it can be further seen that vehicles equipped with a catalytic converter produce

Table 8. Nitroarenes (μg/g particle) and direct mutagenicity in particle emissions [18, 23]

Source	Particle-emission mg/mile	1-Nitropyrene μg/g	μg/mile	6-NitroBaP μg/g	μg/mile	Mutagenic activity rev/μg	rev/mile
Air	—	0.18	—	0.9	—	0.15	—
		0.56		2.5		0.6	
Forest fire	—	0.11	—	0.12	—	0.26	—
Motor vehicle exhaust gases:							
Otto engine (cylinder capacity L):							
– leaded (5.7L)	44.5	3.9	0.17	33	1.5	2.2	98,000
– unleaded (4.3L)	23.3	4.3	0.10	17	0.4	1.0	24,000
– catalytic							
converter (2.5L)	38.2	0.6	0.02	0.2	0.01	0.2	7,000
Diesel (2.1L)	306	3.9	1.2	2.3	0.7	0.5	150,000
Diesel (5.7L)	718	8.2	5.9	1.8	1.3	0.4	284,000
Diesel (4.3L)	867	—	—	—	—	0.46	399,000
Diesel (5.7L)*	397	8.0	3.2	< 0.4	< 0.1	0.93	369,000
Diesel (4.3L)*	216	24.5	5.3	1.1	0.23	—	—
Diesel (4.3L)*	307	7.6	2.3	0.2	0.05	0.21	65,000

* = with exhaust gas recirculation; rev = number of revertants; L = liter

substantially less nitro-PAH. This is also valid for other polycyclic aromatic compounds, e.g. the carcinogen benzo[a]pyrene.

In airborne particles from various locations, the following concentrations of some nitro-PAH were found [25].

Table 9. Nitro-PAH concentrations in atmospheric aerosol from various locations in the USA and Norway [25]. Concentrations in $\mu g/g$

Nitro-PAH	Claremont	St. Louis	Washington	Aurskog (Norway)
2-Nitrofluoranthene	2.8	0.56	0.60	0.56
1-Nitropyrene	0.36	0.16	0.20	0.15
2-Nitropyrene	0.08	0.06	0.05	0.17

In the particle and gaseous constituents of Diesel exhaust gases, the following concentrations were found.

Table 10. Concentrations of nitro-PAH in Diesel exhaust gases [66]

Compound	Particle ($\mu g/g$)	Gaseous ($\mu g/Nm^3$)
1-Nitronaphthalene	2.0	0.2
x-Methyl-1-nitronaphthalene	0.7	0.4
2-Nitronaphthalene	2.5	0.6
x-Nitrofluorene	1.0	2.0
2-Nitrofluorene	0.1	1.5
9-Nitroanthracene	0.2	1.0
x-Nitroanthracene	0.5	0.2
Dinitronaphthalenes	1.0	0.4
Dimethyl-nitronaphthalenes	0.6	Traces
1-Nitropyrene	7.0	n.d.
2,7-Dinitrofluorene	0.3	n.d.
Methyl-nitropyrenes	1.4	n.d.
x-Nitrobenzo[a]pyrene	0.3	n.d.
x,y-Dinitropyrene	0.4	n.d.

n.d. = not detected

Some of the mutagenic, particle-bound components in emissions from incomplete combustion processes have been identified [67]. It can be seen that dinitropyrene causes between 20 and 40% of the mutagenicity (Table 11).

5 Biotransformation and Biological Effects

The most important toxicological feature of nitro-PAH is their mutagenicity: they are direct mutagens in the Ames-tests. In cells from mammals they induce point mutation, breaking of the cromatid thread, chromosomal aberrations, and

Table 11. Concentrations of mutagenic components in exhaust gases from diesel, kerosene heaters, fuel gas, and liquified petroleum gas (LPG) [67]

Mutagen	Diesel engine[a]	Kerosene heater[a]	Fuel gas	LPG[b]
1-Nitropyrene	70.7		0.082	0.10
1,3-Dinitropyrene		0.53 ± 0.59	0.028	0.23
1,6-Dinitropyrene	1.2	3.25 ± 0.63	0.36	2.14
1,8-Dinitropyrene	3.4		0.030	1.64
8-Nitrofluoranthene		0.171		
Benzo[a]pyrene	0.06	0.31	16.11	n.d.
Benz[a]anthene			26.3	59.98
Benzo[k]fluoranthene	0.24	0.46	13.50	27.16
Benzo[b]fluoranthene	0.96	0.74		

[a](ng/mg raw extract); [b](ng/mg particle); n.d. = not detected

sister chromatid exchange in mammalian cell cultures. In animal experiments, carcinomas appear at the injection point after subcutaneous injections [68]. The first nitro-PAH that was investigated in the Ames-test was 5-nitroacenaphthene which is an intermediate product from the manufacture of a whitening agent [69]. Later individual nitro-PAH were comprehensively tested for mutagenicity.

5.1 Metabolism

It is more difficult to give generalized statements about the metabolism of nitro-PAH than with many other groups of substances. Metabolism appears to be specific to the substance, species, and tissue and furthermore depends on the physiological condition of the experimental animal. It has been argued that there is a metabolic activation to amino and acetylamino derivatives by bacterial reductases as the nitro-PAH without previous activation become mutagenic by metabolizing cell fraction (S9). The mutagenic effect in bacterial strains that have no reductase activity is considerably reduced. Besides the reduction of the nitro group, a ring oxidation takes place via epoxides to phenols with the OH-group preferentially positioned at C^6 and C^8 (less frequently at C^3). The phenols and diphenols so produced are also present as the amino derivatives but in concentrations several orders of magnitude less. With this and the nature of their activity in mind they can be disregarded in any discussion of gene toxicity [57].

From metabolic studies of drugs it is known that the reduction of nitro groups can result from the action of various enzymes. In mammals a microbial, cytosolic, and microsomal nitro reductase has been identified. The involvement of xanthine oxidase and cytochrome P450 is subject to debate. The respective nitroso compounds and hydroxylamines occur as intermediate stages on the route to the amines.

Whether a nitro-PAH is mutagenic depends on the position of the nitro group on the aromatic ring. If the nitro group is localized on a L-region of the aromatic ring system it arranges itself almost vertically to the ring system to minimize steric interaction. Nitro-PAH with this type of substitution are only slightly or not at all mutagenic [70]. This is because they, due to steric hindrance, cannot bind to the active centre of the nitro reductases, or are reconverted to the original substance by oxygen-depending redox cycles [70]. Further details with reference to the influence of steric factors can be found in the literature [70]. Table 12 shows the mutagenicity of 1-nitropyrene and its derivatives in *Salmonella typhimurium* TA98 and the metabolic transformation in human hepatoma cells (Hep G2). One can see a) a reduction of the nitro group as apposed to ring oxidation, b) a formation of 6- and 8-phenols rather than 2-phenols, and c) hardly any glucuronides are formed [57].

During the metabolic activation, nitrenium ions are formed which irreversibly bind to the bacterial DNA [29]. This hypothesis is supported by the greatly reduced sensitivity of the nitro reductase deficient strains to nitro-pyrenes. The main adduct of the binding of 1-nitropyrene to DNA has been identified as N-(desoxyguanosine-8-yl)-1-aminopyrene, whose concentration correlates positively with the mutagenic effect of 1-nitropyrene on *Salmonella* strains [29]. Figure 9 shows the reductive activation of nitro-PAH by bacteria.

S. typhimurium metabolizes incubated 1-nitropyrene to 1-aminopyrene as the main product and 1-acetyl-aminopyrene as byproduct [71, 72]. N-hydroxy-1-aminopyrene can be formed from 1-aminopyrene and this binds directly on the DNA and can initiate mutations in *S. typhimurium* [73]. The reduction of 1-nitropyrene to N-hydroxy-1-aminopyrene seems to be a decisive step to mutation induction.

Table 12. Mutagenicity of 1-nitropyrene and its derivatives in *Salmonella typhimurium* TA98 and metabolism in human hepatoma cells [57, 70]

Substance	Metabolized quantities (nmol)[a]	TA98[b]
Pyrene		0
1-Nitropyrene	0.12	721
1,8-Dinitropyrene		157,005
1-Nitrosopyrene		4,146
1-Nitropyrene-3-ol	0.04	713
1-Nitropyrene-6-ol		6.3
1-Nitropyrene-8-ol		3.4
1-Aminopyrene	1.12	
1-NP-*trans*-4,5-dihydrodiol	0.02	
1-Nitropyrene-6&8-ol	0.03	
Other organic soluble metabolites	0.72	
Σ 1-Nitropyrene-ol-*O*-sulfates	0.95	
1-Aminopyrene-*N*-glucuronid	0.10	
Other polar metabolites/conjugates	9.01	

[a] 4 μM ^{3}H-nitropyrene was introduced: 24 h incubation period; [b] Revertants per nmol

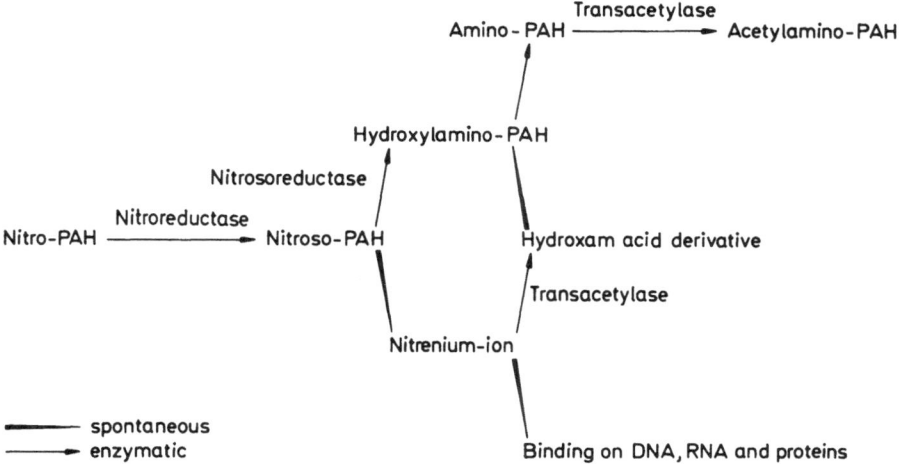

Fig. 9. Reductive activation of nitro-PAH by bacteria [29]

Liver enzymes seem to have special significance in the metabolism of nitro-PAH in mammals [14]. The metabolism of 1-nitropyrene in mammals serves as an example. Under anaerobic conditions, 1-nitropyrene is mainly broken down by liver S9mix (i.e. by activating enzymes, such as monooxygenases, and cofactors), cytosol, or microsomes to 1-aminopyrene [74]. The reaction requires FMN (flavinmononucleotide) and NADPH (reduced nicotinamide-adenine-dinucleotide phosphate) as cofactors [75]. The microsomal reduction of 1-nitropyrene is inhibited by oxygen [76]. The oxidative metabolism of 1-nitropyrene in the rat liver gives rise to 6- and 8-hydroxy-1-nitropyrenes and 1-nitropyrene-*trans*-4,5-dihydrodiol [77]. The possible reduction- and oxidation transformation products from 1-nitropyrene in cytosol and in microsomes are shown in Fig. 10. Very good summaries on the chemical mechanisms of metabolic activation can be found in Beland et al., 1985 [14] and Rosenkranz and Mermelstein, 1983 [3].

5.2 Genotoxicity

As already stated, nitro-PAH in the Ames-test are direct mutagens (i.e. they act without metabolic activation) with *Salmonella typhimurium* [78]. The Ames-test is based on the principle that mutated bacteria of the species *Salmonella typhimurium* LT-2 (histidine dependent) and through the influence of chemicals (or other influences, e.g. radiation) a certain percentage undergo a back muta-tion to the wild type (histidine independent). The strains TA 1535 and TA 1538 mutate back to their wild form and are therefore again capable of synthesizing

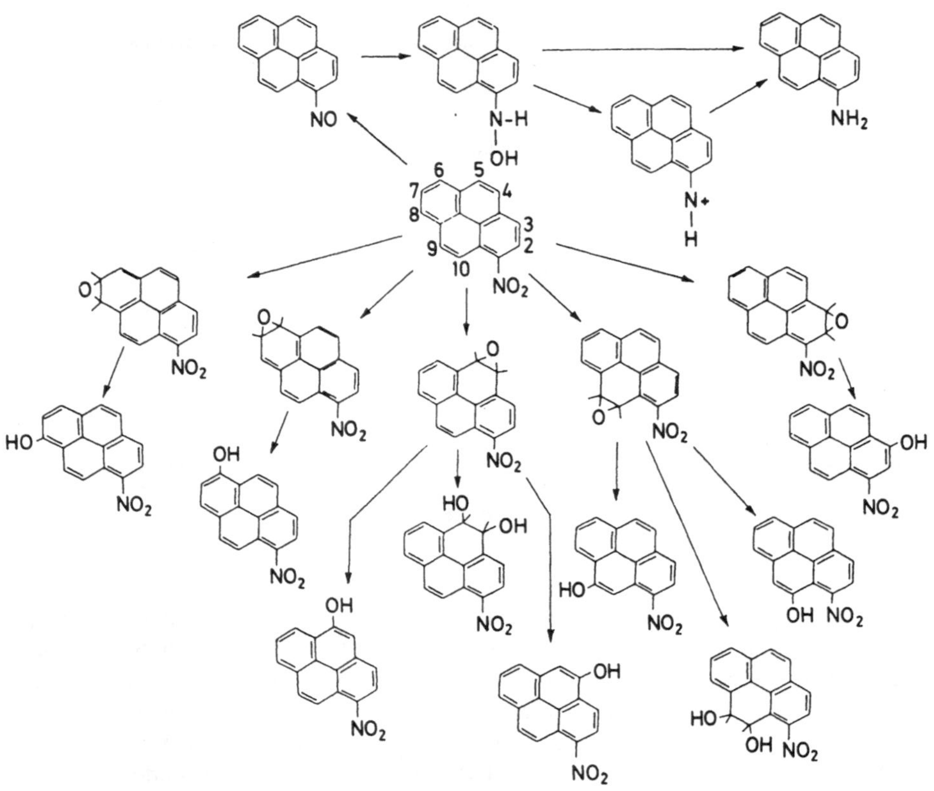

Fig. 10. In vivo and in vitro metabolic transformation of 1-nitropyrene [57]

the histidine necessary for reproduction. Growth of the wild form on a histidine-free culture medium indicates a mutagenic effect. On the addition of the S9mix of enzyme-induced liver homogenates, the formation of the genotoxic products by oxidative metabolic pathways is simulated in mammals, microorganisms, and in cell and tissue cultures. The investigation of more than 300 chemicals which had proven as carcinogenics in animal tests, showed a 90% agreement of carcinogenicity and mutagenicity in the *Salmonella*-bacteria test [79]. Other investigations did not give such a good correlation.

Apart from the direct mutagenicity in the *Salmonella-typhimurium*-test, nitro-PAH induce mutations in bacterial and eucaryotic cells [81]. The mutagenic effect is caused by the reduction of the nitro group [29]. Single strands of DNA can also be broken [82] and sister chromatid exchange processes can be caused [83].

Investigations with 1,8-dinitropyrene show 1-amino-8-nitropyrene as the main metabolite with the twice reduced 1,8-diaminopyrene as the end product [84]. From Table 14 one can see that both of these metabolites (without the liver

Table 13. Genotoxic effect of nitropyrenes on cultured bacteria and mammalian cells [80]

	Test system	Remarks
1-Nitropyrene	S. typhimurium	
1-Nitropyrene 1,3-Dinitropyrene 1,6-Dinitropyrene 1,8-Dinitropyrene 1,3,6-Trinitropyrene 1,3,6,8-Tetranitropyrene	S. typhimurium	
1,3-Dinitropyrene 1,6-Dinitropyrene 1,8-Dinitropyrene 1,3,6-Trinitropyrene	Lung cells of the Chinese hamster Point mutations	
1-Nitropyrene 1,8-Dinitropyrene	Ovary cells of the Chinese hamster	Increase of SCE due to S9
1,8-Dinitropyrene	Mouse lymphoma cells (L 5178 Y) Point mutations	
1,6-Dinitropyrene 1,8-Dinitropyrene	Rat epithelian cells (RL 4) Chromosomal aberrations	
Nitropyrenes, not identified	S. typhimurium, DNA-repair in HeLa-cell cultures	From incubating pyrene with nitro derivatives formed with HNO_3

Table 14. Mutagenicity of 1,8-dinitropyrene and reduction products [84]

S. typhimurum strains	Revertants/nmol (without S9)		
	1,8-Dinitropyrene	1-Aminonitropyrene	1,8-Diaminopyrene
TA-98	92,000	266	15
TA-1538	36,000	98	4.5
TA-1537	32,000	118	4.2
TA-1535		0	0

S9mix) show a mutagenicity lower by two orders of magnitude in the Ames-test. This proves that the mutagenic potential of 1,8-dinitropyrene becomes effective before the first reduction step.

1-Nitropyrene and the 1,3-, 1,6-, and 1,8-substituted dinitropyrenes form, *in vivo* and *in vitro*, metabolites which bind covalently on DNA to give the so-called DNA adducts [85–88]. It is possible that the genotoxic effect manifests itself only after the formation of these DNA adducts.

Some of the nitro-PAH(1,3-, 1,6-, and 1,8-dinitropyrene) belong to the most actively mutagenic substances in the *S. typhimurium* test. Especially the compound with highest biological activity, 1,8-dinitropyrene, is often referred to as being a "super mutagen". The high mutagenicity of 1-nitropyrene, 1,3-, 1,6-, and

1,8-dinitropyrene has been known for a long time. However, there is a series of other nitroarenes which possess a comparable mutagenic potential [53]. Dinitro-PAH, in general, appear to be stronger mutagens than mononitro-PAH [14]. With regard to the mutagenicity potential, great differences can occur between individual nitro-PAH isomers – so, for example, 6-nitrobenzo[a]pyrene is not mutagenic whereas 1- and 3-nitrobenzo[a]pyrenes are very strongly mutagenic substances [14].

In Diesel exhaust gases directly acting mutagenic substances predominate over those which require an initial metabolic activation. The mutagenicity of exhaust gases from gasoline engines and emissions from stationary sources, as shown in the Ames-test, reach their full effect only after having first been activated by a liver S9mix, i.e. by activating enzymes (monooxygenases) and cofactors [89].

The discrepancy between the strong mutagenic effect of nitro-PAH in the Ames-test and the rather weak carcinogenic effect in animal experiments cannot yet be satisfactorily explained [29].

Table 15. Mutagenicity of nitropyrenes and their related compounds [53]

Substance	Mutagenic activity[a]			
	TA100		TA98	
	− S9	+ S9	− S9	+ S9
1,8-Dinitropyrene	14,000	100	400,000	1,300
			250,000	
			200,000	
1,6-Dinitropyrene	13,000	47	160,000	330
			180,000	
1,3-Dinitropyrene	4,200	0	140,000	86
			140,000	
			54,000	
1,3,6-Trinitropyrene	5,000	40	120,000	440
			41,000	
1,7-Dinitropyrene	3,100	77	18,000	260
1,3,6,8-Tetranitropyrene	2,300	23	11,000	110
			16,000	
2,7-Dinitropyrene	68	34	4,500	64
4,10-Dinitropyrene	2,700	0	1,900	140
4,9-Dinitropyrene	1,400	72	680	100
2-Nitropyrene	320	190	1,100	65
	740		2,200	
4-Nitropyrene	760	66	850	34
4,5-Dinitropyrene	460	0	540	0
1-Nitropyrene	190	24	510	96
			470	
			450	
1,7-Dinitro-4,5,9,10-tetrahydropyrene	210	120	98	260
2,7-Dinitro-4,5,9,10-tetrahydropyrene	70	0	86	34
2-Nitro-4,5,9,10-tetrahydropyrene	61	87	18	49
4,9-Dinitrohexahydropyrene	39	18	70	42
4,10-Dinitrohexahydropyrene	19	4.5	35	5.7
4,5-Dinitrohexahydropyrene	0	0	0	10

[a] Revertants per nmol

From the above mentioned substances, 1,8-, 1,6-, and 1,3-dinitropyrene have proved to be carcinogenic in animal experiments; 1-nitropyrene is suspected. Table 15 shows the mutagenicity of nitropyrenes and related substances.

The results shown in Table 15 are from tests with individual substances. In Diesel emissions, however, and in air, particle-bound nitro-PAH and PAH exist together and in mixtures with other substances. In general, nitroarenes are electron acceptors and PAH electron donors. One can therefore conclude that the substances may interact with each other to form charge-transfer complexes. This charge-transfer complex formation may modify the mutagenicity of nitroarenes. An example for this is given in Fig. 11: the direct-acting mutagenic activity of 1,6-dinitropyrene (1-DNP) to T98 decreases in the coexistence of PAH; the reduction is greater the more condensed rings the PAH molecule contains [53].

Futhermore, the direct mutagenicity of nitro-PAH is influenced, i.e. reduced, not only in the presence of PAH but also by other components of the Diesel extract. The presence of Diesel extracts enhances the mutagenic activity of mononitro- and dinitro-PAH. This effect is not so strongly affected by the basic fraction as by the neutral fraction [53].

Early indications of carcinogenic substances were given by experiments on rats: subcutaneous injections of 1-nitropyrene led to malignant fibrous histiocytomas, with one animal developing osteosarcoma [90]. In mice, the subcutaneous injection of 1,6- and 1,8-dinitropyrene also caused malignant fibrous histiocytomas but with 1,3-dinitropyrene, in contrast, no tumor formation was

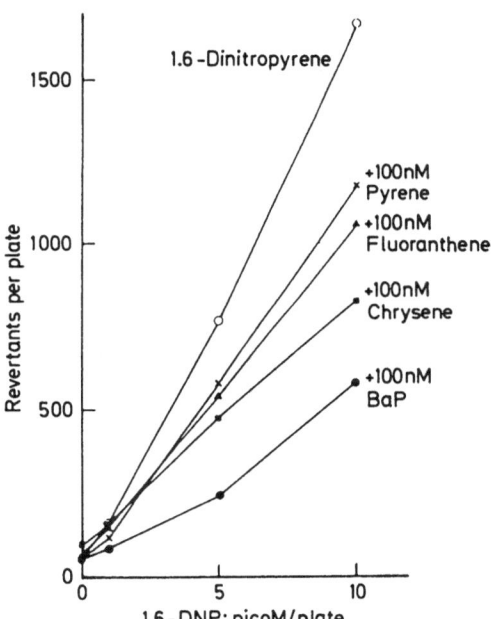

Fig. 11. Reduction in the direct mutagenicity (T98 without S9mix) of 1,6-dinitropyrene in the presence of various PAH [53]

observed [91, 92]. Investigations carried out with rats and hamsters led to sarcomas appearing at the injection site; metastases did not appear [93]. In new-born mice, further nitro-PAH were identified as being carcinogenic [94].

It has been concluded from all the investigations known up to the present time that tumor induction by nitropyrenes proceeds analogously to that of the carcinogenic amines [95]. N-substituted aryl compounds cause tumors and mutations by the reaction sequence:

1) Transformation to N-hydroxy derivatives,
2) Esterification of the N-hydroxy metabolites, and
3) Reaction with nucleic acids.

5.3 Risk Assessment

An evaluation of these findings with reference to the exposure of people to nitro-PAH from combustion processes is at present an extremely difficult task. Detailed investigations into metabolism and carcinogenicity are lacking and these are necessary to judge the relevance to inhaling these substances. Of paramount importance is pollution by particles. Knowledge accumulated until now of the potential mutagenicity and carcinogenicity of Diesel exhaust components results from exposing experimental animals to concentration orders of magnitude greater than a person normally would be [96]. This means that, currently, we cannot definitively answer the question whether gasoline engine emissions expose us to a risk the same as, higher than, or lower than Diesel engine emissions with regard to cancer [80]. Neither is it possible to make a statement on how the effect of nitro-PAH in relationship to the main amount of PAH can be classified and how the initiating effect from chemical carcinogens can be related to the debated (allegedly dose-dependent) particle effect (cf. [5]).

6 Environmental Fate and Photochemistry

During their transport in the atmosphere, organic pollutants undergo chemical transformation in the presence of light, oxygen, water, and an abundance of other noxious substances. The most important transformation of nitro-PAH is photodecomposition. Individual representatives of nitro-PAH, however, are of varying photostability [97]. During photodecomposition, the nitro-PAH is transformed to the respective arylnitrite followed by the formation of the phenoxyl radical with the elimination of nitrogen monooxide [98]. It further reacts to nitrohydroxy derivatives and quinones. Although, the arylnitrite can also hydrolyze to hydroxy derivatives and nitric acid.

Results from Pitts et al. [99] (1978) allow us to assume that nitro-PAH are also denitrated in the atmosphere to the respective quinones. The reaction mechanism might be similar to the one found by Chapman et al. [100] (1966) for 9-nitroanthracene in solution.

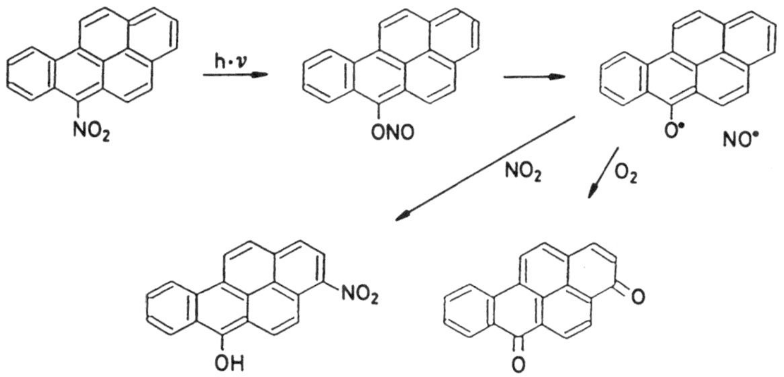

Fig. 12. Photoreaction routes of nitro-PAH using, as example, 6-nitrobenzo[a]pyrene to dioxo- and nitrohydroxy derivatives via aryloxy radicals (adapted from [98], cf. Fig. 6)

The appearance of polycyclic quinones in polluted atmospheres might accordingly be traced back to the photooxidation of PAH and their derivatives. Besides the concentration of PAH and their atmospheric contamination, there are other complex factors to be considered in photooxidation processes e.g. surface chemistry and aerometry (particle size, sunlight intensity, relative humidity, atmospheric mixing, and transport times).

7 Analysis

7.1 Sampling

The concentration of the pollutants on the particles is strongly influenced by the sampling parameters during the separation of the particles from the diluted exhaust gases by filtration, e.g. the type of engine or degree of dilution but also the sampling temperature. According to the definition of the US-EPA particles are all those components, which, at a maximum temperature of 51.7 °C at the sampling probe, are separated on to a defined filter from the air-diluted exhaust gases [48]. Theoretical estimates and experimental results show good agreement, as in diluted Diesel exhaust gas (typical particle concentration 10^{-8} g/cm^3) at temperatures of less than 51.7 °C, more than 95% of the hydrocarbons with a molecular weight greater than 200 are adsorbed on the particles. At lower particle concentrations the adsorption becomes increasingly incomplete. This means that by sampling only particles we can no longer be sure that compounds, e.g. PAH, that occur in a concentration of approximately 10^{-10} g/cm^3 in Otto engine exhaust gases can be quantitatively collected [47].

The fundamental condition for the exact analysis for environmental chemicals in complex sample matrices is a sampling technique from which the collected sample material remains unaffected both qualitatively and quantitatively. To guarantee exactness, sampling particularly when identifying engine emissions must be done with great care.

Two main methods are available for the investigation of nitro-PAH in exhaust gases from combustion engines.

1) *Direct-sampling method* (*total exhaust volume*): By this method [101, 102] all the exhaust gases produced are collected by condensation and filtration.
2) *Dilution tunnel method*: This is the method described by the US Environmental Protection Agency (EPA) for the analysis of the particle emission from Diesel engines. The particle-adsorbed nitro-PAH are filtered out of a small portion of a total flow that has previously been diluted with filtered fresh air. Results obtained with this method can be found in Hartung et al. [103] (1982) and Lies et al. [104] (1986). The formulation of VDI (Association of German Engineers) regulation also envisages the use of the dilution method [105, 106].

The dilution method is the more flexible method since, besides the exhaust gas also the particulates and other exhaust components are sampled [107]. The principle is shown in Fig. 13.

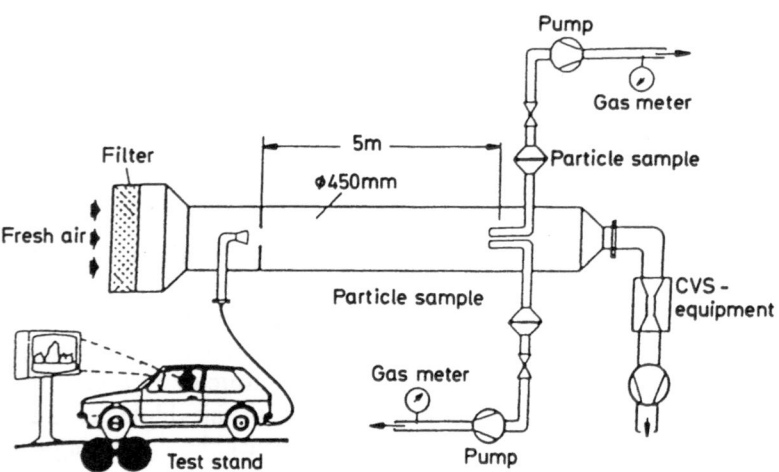

Fig. 13. Test installation for the dilution method of collecting particles from Diesel exhaust gas [46]

A sampling device which enables the measurement of pollutants directly from the combustion chamber has been described [108]. The principle is that a valve is introduced into the combustion chamber and aliquot portions at the high combustion temperatures (under pressure) are taken. This device made it finally possible to find out if nitro-PAH are formed in the combustion chamber or later in the exhaust system. The values of the PAH-measurements in the combustion chamber were 2–3 orders of magnitude greater than in the exhaust gases leaving the exhaust pipe; the mutagenic activity in the exhaust particles was one order of magnitude less than in the combustion chamber.

7.2 The Problem of Artifact Formation

Pure, dissolved PAH react with NO_2 on the filter material due to electrophilic aromatic reactions to the corresponding nitro derivatives [99, 109]. If this was true for PAH adsorbed on particles, then the concentrations of nitro-PAH in exhaust gases from engines would be overestimated especially for longer collection periods. The first evidence for such artifacts, which had previously been put forward by Gibson et al. [110] (1981), has been proved by reliable and extensive surveys [46, 111]. For sampling periods of less than 45 minutes under conditions corresponding to US regulations (Federal Test Procedure), the artifact formation for 1-nitropyrene was less than 20% of the 1-nitropyrene adsorbed on particles.

Figure 14 shows in detail a measuring apparatus which enables the investigation of artifact formation to be carried out.

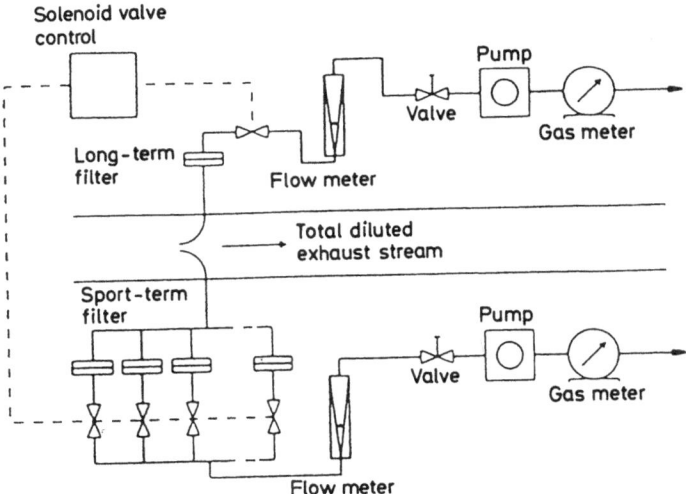

Fig. 14. A diagrammatic representation of a sampling system for the investigation of artifact formation [46]

The primary source of 1-nitropyrene in Diesel exhaust gas has been thus shown to be the formation in the engine and exhaust system. It can be assumed that it is similar for other nitro-PAH.

The following equation describes the formation rate of nitro-PAH [46]. This enables us to estimate the amount of artifact formation:

$$\frac{d[\text{nitro-PAH}]}{dt} = k \times [\text{PAH}] \times [\text{NO}_2]$$

$[X]$ = concentration of X (ppm)

k = rate constant $(h^{-1} \times ppm^{-1})$

A rate constant for 1-nitropyrene has already been determined: $k = 1.4 \times 10^{-2} h^{-1} \times ppm^{-1}$ [109]. No temperature dependence of k was observed between 34 °C and 52 °C. For some nitro-PAH (nitrofluoranthene, nitrobenzofluoranthene, nitroindeno[1,2,3-cd]pyrene) the formation rate can be very small and artifact formation for normal collection periods of up to 90 minutes can be disregarded. This is the result of the low reactivity of the initial PAH in electrophilic aromatic reactions. Table 16 shows a list of PAH, normally found in Diesel exhaust gases, classified according to their reactivity in electrophilic aromatic reactions.

The reactivity of the PAH formed in the engine is directly proportional to the temperature in the sampling system. This means that to quantitatively determine the nitro-PAH formed in the combustion processes the hot exhaust gases must be immediately cooled on leaving the exhaust system. In this way, the electrophilic reactions of the PAH with reactive gases, which lead to artifact formation, can be reduced to a minimum. The incorporation of devices to separate off these reactive gases such as HCl, NO_x, and SO_2 from the gas phase is worth considering. For the analysis of refuse incineration plant emissions, this type of apparatus is already in use [113]. It is known as the PUFP-method (*Polyurethane Foam Plug*).

To ensure reliable and accurate results, besides these details of the construction of the apparatus it is essential that there is a continuous determination of the recovery of the substances and regular checking of the effectivity and capacity of the filter and adsorption material.

Table 16. Classification of PAH-reactivity in electrophilic aromatic reactions in order of diminishing reactivity [112]

Group 1:	Benzo[a]tetracene, pentacene, tetracene
Group 2:	Anthracene, anthanthrene, benzo[a]pyrene, perylene
Group 3:	Benz[a]anthracene, benzo[ghi]perylene, cyclopenteno[c,d]pyrene, pyrene
Group 4:	Benzo[c]phenanthrene, benzo[e]pyrene, chrysene, coronen, dibenzanthracenes
Group 5:	Benzofluoranthenes, fluoranthene, triphenylene, indeno[1,2,3-c,d]pyrene, naphthalene, phenanthrene

7.3 Analytical Workup of Environmental Samples

Complex mixtures of substances cannot be separated efficiently by chromatographic methods without previous concentration and fractionation.

7.3.1 Extraction

The determination of the nitro-PAH begins with their extraction from the sample matrix and then concentration of the extracting agent by rotation evaporation. The extraction is carried out by boiling of the sample with a solvent in the Soxhlet apparatus or by treatment in an ultrasonic bath. For quantitation of the recovery an internal standard is added to the sample. This should behave in a similar way to the substances being investigated allowing conclusive statements about the quality of the extraction and the complete treatment of the samples.

To avoid overlapping, the internal standard should not already be contained *a priori* in the Diesel emissions. For nitro-PAH, 3-nitrofluoranthene would be an ideal compound with the advantage that its high photosensitivity, furthermore, allows the quantification due to losses through photodecomposition.

The Soxhlet method gives only marginally better yields than extraction with an ultrasonic apparatus [114]. Soxhlet extraction ensures that 99% of the total nitro-PAH is extracted in 8 hours [114]. However, extraction with ultrasound needs simpler apparatus and can also be carried out at higher temperatures than the customary room temperature.

Binary solvent systems are generally used for extraction. Such mixtures consisted either an aromatic component and an alcohol (e.g. toluene/propanol = 51/49) or aromatic solvents in combination with dichloromethane (e.g. toluene/dichloromethane = 3/1) [115]. The extract is often filtered through a Teflon membrane before being concentrated by the rotatory evaporator [115, 116].

7.3.2 Fractionation

The chemical composition of the particles, e.g. in Diesel exhaust gases, is extremely complex. The organic extracts consist of aliphatic hydrocarbons, PAH, nitro-PAH, N- and S-heterocycles, and a multitude of other organic compounds. To analyze single classes of compounds or individual species, it is necessary for the solvent extract to undergo fractionation. This can be done chromatographically in florisil- or silica gel columns with eluants of varying polarity [115, 116].

Another method of fractionating is the preparative separation by high-pressure liquid chromatography (HPLC). By suitable selection of the column

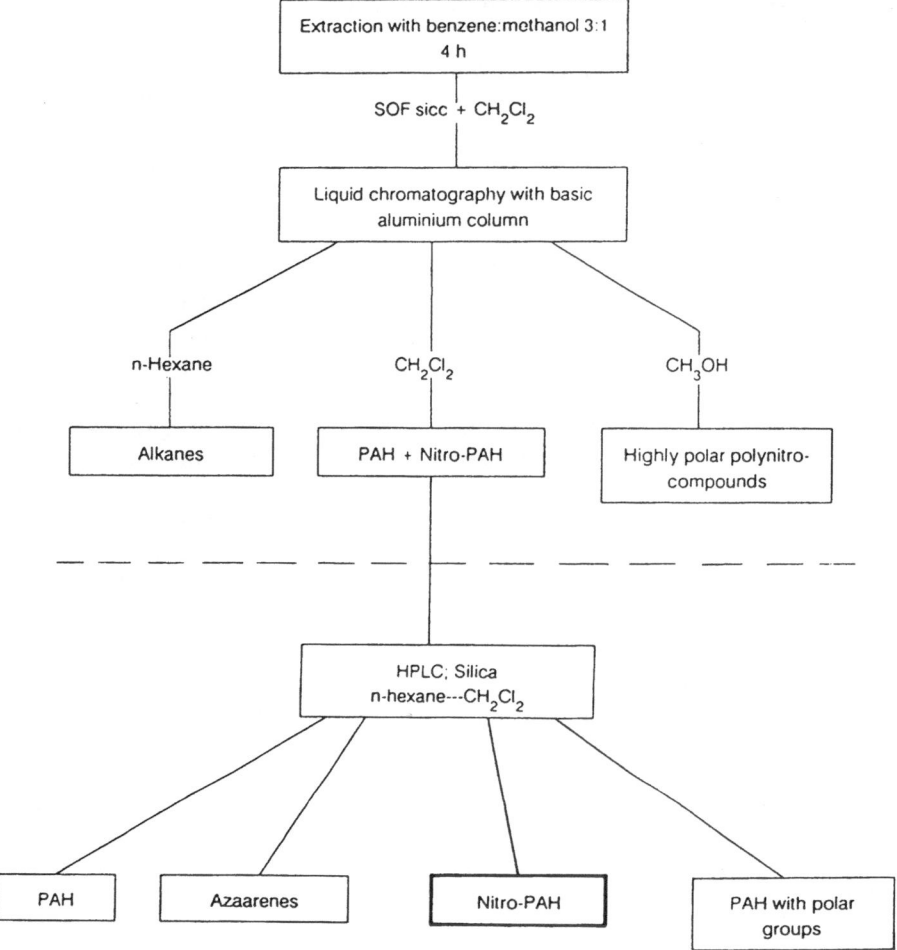

Fig. 15. Fractionation and purification of extracts from particle and gaseous emissions from Diesel engines for the separation of nitro-PAH by liquid chromatography [66]

material and use of pressure- and/or eluant gradient programs, comprehensive separation of the extract can be achieved.

The concentrated eluate, can be used for the final analytical determination. For this, there are many possibilities, some of which are not suitable for routine investigations due to factors such as high costs or the necessity for specially trained personell. A possible fractionating scheme is shown in Fig. 15.

7.3.3 Detection

The last step in analysis consists in the chromatographic separation and the detection of the compounds. The chromatographic methods available are: gas chromatography, high pressure liquid chromatography, and high performance thin layer chromatography. The detection methods include nitrogen selective detectors (NSD), electron capture detectors (ECD), flame ionization detectors (FID), thermal conductivity detectors (TCD), mass-spectrometric methods (MS), and detectors which work on the principles of fluorescence, chemoluminescence, or electrochemical methods. Individual methods naturally vary in respect to their selectivity and sensitivity. The respective methods have to be selected and tested according to the requirements.

A further refinement of the chromatographic separation and following detection is to connect in series the same or differing chromatographic methods and suitable combinations of several detectors. These are known as multidimensional methods.

Figure 16 shows a typical gas chromatogram of a Diesel extract with a nitrogen selective detector, Fig. 17 a HPLC-chromatogram with fluorescence (upper part) and UV-detection (lower part), and Fig. 18 a gas chromatogram of a Diesel extract after fractionating by HPLC.

Refining the methods makes it possible to reliably detect and quantitate increasingly lower amounts of nitro-PAH. With a multidimensional gas chromatography systems and nitrogen selective detection it was possible to achieve a detection limit of 36 pg (picogram = 10^{-12} g) which, in addition was highly reproducible [115]. With the aid of multidimensional HPLC followed by fluorescence detection for 1-nitropyrene after its reduction to the respective amine it was even possible to lower this detection limit to 10 pg [117].

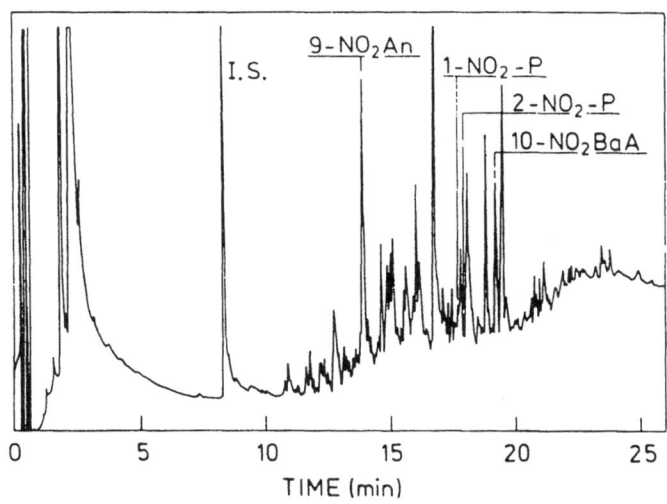

Fig. 16. Gas chromatogram of a Diesel extract detected with a nitrogen selective detector (NSD) after fractionating by HPLC [66]

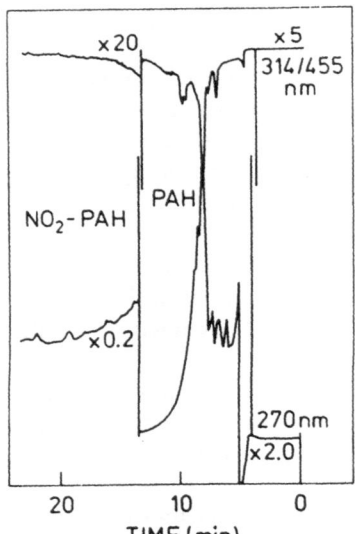

Fig. 17. HPLC-chromatogram of a sample containing PAH and nitro-PAH. Detection above: UV-absorption; below: fluorescence [66]

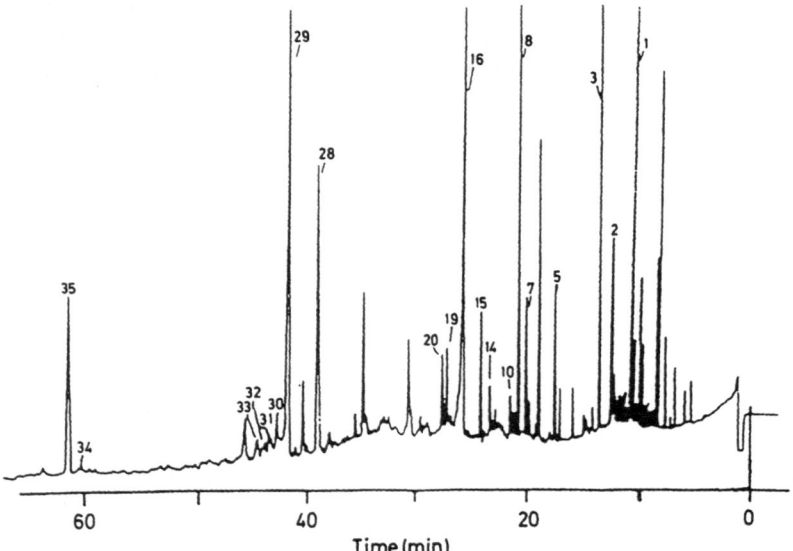

Fig. 18. Gas chromatogram of a Diesel extract detected with a nitrogen selective detector (NSD) after fractionation by HPLC [66]

1 = 1-Nitronaphthalene; 2 = x-Methyl-1-nitronaphthalene; 3 = 2-Nitronaphthalene;
5 = x-Methylnitrobiphenyl; 7 = x-Nitrofluorene; 8 = x,y-Dinitrobiphenyl;
10 = 2-Nitrofluorene; 14 = 9-Nitroanthracene; 15 = x-Nitroanthracene;
16 = x,y-Dinitronaphthalene; 19 = x,y-Dimethylnitronaphthalene;
20 = x,y-Dimethylnitronaphthalene; 28 = 1-Nitropyrene; 29 = 2,7-Dinitrofluorene;
30 = x-Methylnitropyrene; 31 = x-Methylnitrofluoranthene; 34 = Nitrobenzo[a]pyrene;
35 = x,y-Dinitropyrene

7.3.4 Identification

The analysis methods for non-limited exhaust gas components are not regulated by law just as it is for sampling techniques. Unregulated emissions are defined as those exhaust gas products for which emission standards according to a driving cycle have not been promulgated. There are three fractions of PAH that have to be worked up: water condensate, coating of the cooler and of the filter. The last usually contains the major proportion of PAH. The PAH extracted from the three fractions are combined and by liquid-liquid partition twice separated from impurities. For nitro-PAH, capillary gas chromatography with nitrogen specific thermo-ionization detection (GC/TID) and high pressure liquid chromatography (HPLC) with on-line reduction of the nitro-PAH to their amino derivatives and finally fluorescence detection has proven useful [111].

For identifying the compounds, a coupling of gas chromatography-mass spectrometry (GC/MS) or mass spectrometry-mass spectrometry (MS/MS) is suitable [118–120]. The disadvantage of the MS/MS coupling is, however, that it cannot discriminate between isomeric compounds. Methods of synthesizing some nitro-PAH which can be used for quantitating the results can be found in Bodine et al. [121] (1981). Selective detection methods for nitro-PAH include the following [122]:

— The nitro group may be chemically reduced to amino group and followed by reaction)
— NO_2-PAH may be reduced electrochemically followed by detection as the corresponding aromatic amine
— Photolytical degradation to nitrite ions followed by electrochemical oxidation and detection as nitrate ions
— Highly electronegative nitro groups may be detected with the electron capture detector or the thermoionic ionization detector operated at moderate temperatures in an inert atmosphere (e.g. nitrogen). The same principle applies for negative-ion chemical ionization mass spectrometry.
— NO_2-PAH may be detected with a nitrogen-phosphorous detector.
— Nitro groups may be decomposed thermally to nitrosyl radicals then reacted with ozone and finally detected as photons (chemiluminescent reaction)
— Use of various mass spectrometric methods, such as selective ionization procedures, selective ion monitoring, or exact mass measurement.

An overview of analytical methods normally used is given in Table 17.

8 Emissions from Combustion Engines

In the flow of exhaust gases from combustion engines, PAH occur mainly particle bound. The majority of these are smaller than 2 μm and can therefore be inhaled into the lungs [135]. Structural restrictions are unlikely to occur in the

Table 17. Analytical methods for determination of nitro-PAH [122]

Chromatographic method and detector	Linear range	Detection limit	Ref.
TLC/Fluorescence	Not given	1–500 ng[a]	[123]
Capillary GC/NPD	Not given	200–500 pg, S/N ~3[b]	[124]
Capillary GC/ECD	1 Order of magnitude	20–60 pg[c]	[125]
Capillary GC/TID-1-N$_2$	3 pg–110 ng	3 pg, S/N ~3[d]	[126]
Capillary GC/TEA	Not given	25–80 pg, S/N ~3[e]	[127]
Packed GC/ECD/PID	Not given	ECD: 3–66 pg[f], PID: 4–162 ng	[128]
Packed GC/TEA	3–4 Orders of magnitude	300–600 pg[g]	[129]
HPLC/Fluorescence	Not given	1–25 pg[h]	[130]
HPLC/MS	Not given	7 ng for NO$_2$-pyrene	[131]
Reductive LC/EC with 25 cm × 4 mm i.d. ODS column	0.1–100 ng	10–100 pg[i]	[132]
Reductive LC/EC with 50 cm × 1 mm i.d. microbore ODS column	1 to 3 Orders of magnitude	0.001–1 ng[j]	[133]
Oxidative LC/EC	2 Orders of magnitude (1–100 ng)	0.5–4.5 ng[k]	[134]

[a] For 1-nitronaphthalene, 9-nitroanthracene, 6-nitrochrysene, 3-nitrofluoranthene, 1-nitropyrene, 6-nitrobenzo[a]pyrene, and 1,6-dinitropyrene; [b] 1-Nitronaphthalene and 1,3-dinitropyrene; [c] For nitrated pyrene and nitrated benzo[a]pyrene; [d] For 2,2′-dinitrobiphenyl and 9-nitroanthracene; [e] For 1-nitronaphthalene, 1,5-dinitronaphthalene, 1,4,5-trinitronaphthalene, 4-nitrobiphenyl, 9-nitroanthracene, 2-nitrofluorene, 2,7-dinitrofluorene, nitrofluoranthene, 1-nitropyrene, and 3-nitropyrene; [f] For a variety of mononitrated and dinitrated benzenes[*]; [g] For 2-nitronaphthalene, 3-nitrobiphenyl, 4-nitrobiphenyl, 2-nitrofluorene, 9-nitroanthracene, and 1-nitropyrene; [h] For various optimized wavelength pairs. For 3-aminofluoranthene, 1-aminopyrene, 6-aminobenzo[a]pyrene, and aminobenzo[k]fluoranthene; [i] Based on 2,7-dinitrofluorene, 2-nitrofluorene, 3-nitro-9-fluorenone, and 1-nitropyrene; [j] Based on 1-nitronaphthalene, 2-nitrofluorene, 9-nitroanthracene, 1-nitropyrene, 4-nitrofluoranthene, 2,7-dinitrofluorene, and 2,7-dinitrofluorene; [k] For RDX, tetryl, TNT, and nitroglycerin

Abbreviations:

EC	Electrochemically	ECD	Electron capture detector
NPD	Nitrogen-phosphorous detector	PID	Photoionization detector
TEA	Thermal energy analyzer		
TID-1-N$_2$	Thermionic ionization detector in the nitrogen mode		

most common incomplete combustion processes. This is due to the fact that the reactions involve short-lived radicals and are therefore mainly diffusion controlled and require no activation energy which is dependent on the initial or end products of the individual steps [136]. Unstable PAH stabilize themselves in subsequent reactions, and disproportionation, intra- and intermolecular ring closure are probably the most common stabilizing reactions.

More than 200 nitro-PAH have been detected in the Diesel exhaust gases. The largest proportion being that of 1-nitropyrene with between 107 and 1590 ppm (in relation to the weight of the extract), followed by nitro-phenanthrene/-anthracene isomers [137]. Dinitropyrene isomers were identified at a level in the sub-ppm region (0.4–0.6 ppm). In spite of the high proportion of

Table 18. The contribution of nitro-PAH to the mutagenicity of particle bound extracts in *Salmonella typhimurium* TA98 (− S9mix) [137]

Exhaust extract sample	Mutagenic activity (Rev/μg) TA98 − S9	1-Nitropyrene (ppm)	(%)	Nitrofluoranthene (ppm)	(%)	Dinitropyrene Isomers (ppm)	(%)
Diesel	13.0	1,590	11.0	7.0	1.4	—	—
	3.9	589	13.0	1.2	0.8	1.6	26.0
	3.5	107	2.7	0.9	0.8	—	—
Gasoline	1.6	2.5	0.1	—	—	—	—

Table 19. Emission values of nitro-PAH derivatives in the exhaust gas of a Diesel vehicle (warm start); concentrations in μg/mile

Substance	4-cylinder-Diesel engine Average of three measurements
2-Nitrofluorene	1.5
2,7-Dinitro-9-fluorene	0.2
9-Nitroanthracene	1.6
3-Nitrofluoranthene	0.1
1-Nitropyrene	5.5
Σ Dinitropyrenes (1,3-, 1,6-, 1,8-)	0.1
6-Nitrobenzo[a]pyrene	0.1

1-nitropyrene in Diesel extract, this substance is only responsible for 3–13% of the mutagenicity. The small (0.4–0.6 ppm) amounts of the 1,3-, 1,6-, and 1,8-substituted dinitropyrenes (total amount = 1.6 ppm) are, in contrast, responsible for approximately 26% of the mutagenicity.

Although there are few facts available on the direct formation of nitro-PAH in the engine in relation to various parameters, it can be assumed that nitro-PAH are formed by nitration reactions by referring to the unequivocal results of PAH formation in the engine.

With the exception of the 1-nitropyrene, the nitro-PAH are in much lower concentrations than the underivatized or oxy-PAH (Table 19).

Although numerous PAH have been detected in automobile exhaust gases, only a few are routinely monitored (this saves a great deal of time required for analysis and makes the clean up simpler) [104]. A detailed study of the formation of PAH in relation to various engine parameters was reported by Lepperhof [138] (1981). Although a reduction in the PAH emission of approximately 80% can be expected [138] by using a 20% leaner mixture (air-fuel ratio $\lambda = 0.9$ to $\lambda = 1.1$), in comparable types of automobiles with gasoline engines, differences up to a factor of four occur have been determined for 1-nitropyrene. Emission rates (μg/min) have been determined to be as follows: Oldsmobile 3.0 $\pm$ 1.0; Mercedes 7.8 $\pm$ 1.2; Volkswagen 1.8 $\pm$ 0.9; Peugeot 3.8 $\pm$ 1.4. The chemical composition of the fuel (% of aromatic compounds) has no influence of the amount of the 1-nitropyrene formed (but for benzopyrenes this is a decisive factor [139]. At the moment no measurements exist determining the influence of the load on the amount nitro-PAH emitted. The ignition timing does not seem to influence the emission rate of 1-nitropyrene [139]:

Standard ignition	0.9 ± 0.02 (μg/min)
Advanced ignition	1.0 ± 0.01 (μg/min)
Retarded ignition	1.3 ± 0.10 (μg/min)

9 Relevance of the PAH-Emission Results to the Emission of Nitro-PAH

There is no first hand knowledge about the direct formation of nitro-PAH in the engine in relation to various parameters. It can be assumed, however, that since nitro-PAH are formed from PAH by nitration reactions, the relevant results from investigations into PAH formation in the engine can be used.

The nitro-PAH are probably not formed in the engine itself, but rather in the subsequent combustion phase with the simultaneous presence of PAH, nitrogen oxides, and traces of acids which act as catalysts [3]. When one assumes that nitro-PAH are formed only by nitration of PAH, the PAH must therefore be formed by radical polymerization processes of short-chained hydrocarbon species. That means that a great part of the reaction time is necessary for the synthesis of the PAH [140]. However, it might be that radical intermediate products are nitrated even before the ring closing [140].

A reduction in nitro-PAH formation therefore seems possible by varying the engine parameters leading to a lower emission of PAH. In order to lower PAH emissions the following measures are available:

— Running engines on predominantly lean mixtures,
— Low gap width between the gap and cylinder wall,
— Raising the cylinder-wall temperatures.

Another alternative would be to reduce the nitro-PAH by measures that reduce the nitrogen oxides essential for nitro-PAH formation [141]. The use of 3-way catalytic converters for automobiles can be expected to lead to a reduction in the nitro derivatives of PAH.

10 References

1. Nielsen T (1981) Nitro derivatives of polynuclear aromatics: Formation, presence and transformation in stack and exhaust gases and in the atmosphere. Riso National Laboratory, DK-4000 Roskilde, Denmark

2. Nielsen T (1984) Environ Sci Technol 18:157
3. Rosenkranz HS, Mermelstein R (1983) Mutat Res 114:217
4. Hirose M, Lee M-S, Wang CY, King CM (1984) Cancer Res 44:1158
5. Mücke W (1988) Das öffentliche Gesundheitswesen 50:147
6. DFG (Deutsche Forschungsgemeinschaft) (1988) Mitteilung XXIV der Senatskommission zur Prüfung gesundheitsschädlicher Arbeitsstoffe: Maximale Arbeitsplatzkonzentrationen und Biologische Arbeitsstofftoleranzwerte. VCH Verlagsgesellschaft, Weinheim
7. Jacob J, Grimmer G, Naujack K-W (1986) Erdöl und Kohle – Erdgas – Petrochemie 39:140
8. Neumüller O-A (1985) Römpps Chemie Lexikon, Band 4, 8. Auflage, Franckh'sche Verlagsbuchhandlung, Stuttgart
9. Ramdahl T (1983) Environ Sci Technol 17:666
10. Fisher GL, Raabe OL, Chrisp CE (1979) Science 204:879
11. Ramdahl T, Becher G, Bjorseth A (1982) Environ Sci Technol 16:861
12. Nielsen T (1983) Anal Chem 55:286
13. El-Bayoumy K, O'Donnell M, Hecht SS, Hoffmann D (1985) Carcinogenesis 6:505
14. Beland FA, Heflich RH, Howard PC, Fu PP (1985) The in vitro metabolic activation of nitro polycyclic aromatic hydrocarbons. In: Harvey RG (ed) Polycyclic hydrocarbons and carcinogenesis. ACS Symposium Series, No. 283, American Chemical Society, Washington DC
15. Schuetzle D, Lee FS-C, Prater TJ, Tejada SB (1981) Int J Environ Anal Chem 9:93
16. Talcott R, Harger W (1980) Mutat Res 79:177
17. Grimmer G (1983) Profile analysis of polycyclic aromatic hydrocarbons in air. In: Björseth A (ed) Handbook of polycyclic aromatic hydrocarbons, Marcel Dekker, New York, p 149
18. Gibson TL (1983) Mutat Res 122:115
19. Hanson RL, Henderson TR, Hobbs CH, Clark CR, Carpenter RL, Dutcher JS, Harvey TM, Hunt DF (1983) J Toxicol Environ Health 11:971
20. Chan TL, Gibson TL (1985) Sampling and atmospheric chemistry of particles containing nitrated polycyclic aromatic hydrocarbons. In: White CM (ed) Nitrated polycyclic aromatic hydrocarbons, Huethig, Heidelberg, p 227
21. Volz A, Ehhalt DH, Derwent RG (1981) J Geophys Res Ser C 86:5163
22. Lewtas J, Nishioka MG, Patterson BA (1982) The role of nitroaromatics in the mutagenicity of environmental emissions. Unpublished results (Citation found. In: Rickert DE et al. (eds) The toxicity of nitroaromatic compounds
23. Gibson TL (1982) Sources of direct-acting nitroarene mutagens in airborne particulate matter. Environmental Science Department, General Motors Research Laboratories, Warren, Michigan 48090, USA
24. Tully FP, Ravishankara AR, Thompson RL, Nicovich JM, Shah RC, Kreutter NM, Wine PH (1981) J Phys Chem 85:2262
25. Ramdahl T, Zielinska B, Arey J, Atkinson R, Winer AM, Pitts JN Jr (1986) Ubiquitous occurrence of 2-nitrofluoranthene and 2-nitropyrene in air. Nature 321:425
26. Arey J, Zielinska B, Atkinson R, Winer AM, Ramdahl T, Pitts JN Jr (1986) The formation of nitro-PAH from the gas phase reactions of fluoranthene and pyrene with the OH radical in the presence of NO$_x$. Atmospheric Environ 20:2339
27. Atkinson R, Carter WPL, Darnall KR, Winer AM, Pitts JN Jr (1980) Int J Chem Kinet 12:779
28. Chuang ChC, Mack GA, Petersen BA, Wilson NK (1984) Identification and quantitation of nitropolynuclear aromatic hydrocarbons in ambient and indoor air particulate samples, p 155. Battelle, Columbus, Ohio
29. Kouros M, Dehnen W (1985) Funkt Biol Med 4:82
30. Platt U, Perner D, Winer AM, Harris GW, Pitts JN Jr (1980) Geophys Res Lett 7:89
31. Pitts JN Jr, Sweetman JA, Zielinska B, Atkinson R, Winen AM, Harger WP (1985) Environ Sci Technol 19:1115
32. Jäger J, Hanus V (1980) J Hyg Epidemiol Microbiol Immunol 24:1
33. Mosberg A, Mays D, Riggin R, Shure M, Mumford J, Fisher G (1981) Chemical and physical properties of vapor phase nitropyrene-coated coal fly ash. In: Sixth International Symposium on Polynuclear Aromatic Hydrocarbons, Abstracts, Battelle Laboratories, Columbus, Ohio, S. 70
34. van Cauwenberghe KA, van Vaeck L, Pitts JN (1979) Adv Mass Spectrom 8(B):1499
35. Butler JD, Crossley P (1981) Atmospheric Environ 15:91
36. Pedersen PS, Ingwersen J, Nielsen T, Larsen E (1980) Environ Sci Technol 14:71
37. Grandjean P, Nielsen T (1979) Residue Rev 72:97
38. Coughlin RW, Ezra FS (1968) Environ Sci Technol 2:291

39. Thomas JF, Mukai M, Tebbens BD (1968) Environ Sci Technol 2:33
40. Olah GA, Narang SC, Olah JA (1981) Proc Nat Acad Sci USA 78:3298
41. Muha GM (1970a) J Phys Chem 74:787
42. Muha GM (1970b) J Phys Chem 74:2939
43. Grosjean D, Fung K, Harrison J (1983) Environ Sci Technol 17:673
44. Nielsen T (1984) Karakterisering af polycyclisk organisk materiale (POM) i roggasser fra kulfyrede forbraendingsanlaeg i atmosfaeren og undersogelse af deres omdannelse i atmosfaeren. Riso National Laboratory, DK-4000 Roskilde, Denmark
45. Sweetman JA, Zielinska B, Atkinson R, Ramdahl T, Winer AM, Pitts JN Jr (1986) Atmospheric Environ 20:235
46. Hartung A, Schulze, Kieß H, Lies K-H (1986) Staub-Reinhalt Luft 46:132
47. Lies K-H, Schulze J, Winneke H, Kuhler M, Kraft J, Hartung A, Postulka A, Gring H, Schröter D (1988) Nicht limitierte Automobil-Abgaskomponenten. Volkswagen AG, Wolfsburg/F.R.G.
48. U.S. EPA (1979) U.S. Air Pollutant Emission Inventory. EPA 450/4-81-010
49. Pott F (1985) Staub-Reinhalt Luft 45:369
50. Sklorz M (1989) Bestimmung von polycyclischen Aromaten und ihrer Nitroderivate an atmosphärischen Partikeln. Masters Thesis, University of Bayreuth, Chair of Ecological Chemistry and Geo-chemistry, F.R.G.
51. Grimmer G (1985) Vorkommen, Analytik und Bedeutung der PAH als Umweltcarcinogene. Erdöl und Kohle – Erdgas – Petrochemie 38:310
52. Israël G, Freise R, Bayer H-W (1985) Verkehrsbeitrag zur Gesamtstaub- und PAH-Immission in deutschen Groß-Städten. Staub-Reinhalt Luft 45:353
53. Matsushita H, Goto S, Endo O, Lee J-H, Kawai A (1986) Mutagenicity of diesel exhaust and related chemicals. In: Ishinishi N, Koizumi A, McClellan RO, Stöber W (eds) Carcinogenic and mutagenic effects of diesel engine exhaust, p 103, Elsevier, Amsterdam
54. Rosenkranz HS, McCoy EC, Sanders DR, Butler M, Kiriazides DK, Mermelstein R (1980) Science 209:1039
55. Rosenkranz HS (1982) Mutat Res 101:1
56. Rosenkranz HS, Mermelstein R (1985) Environ Carcinog Rev C3:221
57. Rosenkranz HS, Howard PC (1986) Structural basis of the activity of nitrated polycyclic aromatic hydrocarbons. In: Ishinishi N, Koizumi A, McClellan RO, Stöber W (eds) Carcinogenic and mutagenic effects of diesel engine exhaust, p 141, Elsevier, Amsterdam
58. Nielsen T, Seitz B, Ramdahl T (1984) Atmospheric Environ 18:2159
59. Yamaki N, Kohno T, Ishiwata S, Matsushita H, Yoshihara K, Iida Y, Mizoguchi T, Okuzawa S, Sakamoto K, Kachi H, Goto S, Sakamoto T, Daishima S (1986) The state of the art on the chemical characterization of diesel particulates in Japan. In: Ishinishi N, Koizumi A, McClellan RO, Stöber W (eds) Carcinogenic and mutagenic effects of diesel engine exhaust, p 17, Elsevier, Amsterdam
60. Gibson TL, Tironi G (1985) Polynuclear aromatic compounds and the bacterial mutagenicity of airborne particulate matter. Polynucl Aromat Hydrocarbons, (Pap Int Symp) 8th, Meeting Date 1983, 463–74. Edited by: Cooke M and Dennis AJ. Battelle Press, Columbus, Ohio
61. Sellström U, Jansson, Bergman A, Alsberg T (1987) Chemosphere 16:945
62. Cuddihy RC, Griffith WC, McClellan RO (1984) Environ Sci Technol 18:14A
63. DeCoufle P, Stanislawczyck K, Houten L (1977) A retrospective survey of cancer in relation to occupation. NIOSH Report No. 77–178
64. Walrath J, Rogot E, Murray J, Blair A (1985) Mortality patterns among US veterans by occupation and smoking status. NIH Publikation Nr. 85–2756; U.S. Government Printing Office, Washington, DC
65. Dubrow R, Wegmann D (1984) Occupational characteristics of cancer victims in Massachusetts, 1971–1973. NIOSH Publication No. 84–109
66. Liberti A, Cicioli P, Cecinato A, Brancaleoni E, Palo CD (1984) J High Res Chromatogr Chromatogr Commun 7:389
67. Tokiwa H, Nakagawa R, Horikawa K, Ohkubo A (1987) Environ Health Perspect 73:191
68. MAK (1983) Maximale Arbeitsplatzkonzentrationen und Biologische Arbeitsstofftoleranzwerte. Mitteilung der Senatskommission zur Prüfung gesundheitsschädlicher Arbeitsstoffe. VCH Verlagsgesellschaft, Weinheim
69. Yahagi T, Shimizu H, Nagao M, Takamura N, Sugimura T (1975) Gann 66:581
70. Fu PP, Chou MW, Miller DW, White GL, Heflich RH, Beland FA (1985) Mutat Res 143:173
71. Messier F, Lu C, Andrews P, McCarry BE, Quilliam MA, McCalla DR (1981) Carcinogenesis 2:1007

72. Howard PC, Heflich RH, Evans FE, Beland FA (1983) Cancer Res 43:2052
73. Beland FA, Beranek DT, Dooley KL, Heflich RH, Kadlubar FF (1983) Environ Health Perspect 49:125
74. Cerniglia CE, Freeman JP, White GL, Heflich RH, Miller DW (1985) Appl Environmental Microbiology 50:649
75. Nachtman JP, Wei ET (1982) Experentia 38:837
76. Saito K, Kamataki T, Kato R (1984) Cancer Res 44:3169
77. El-Bayoumy K, Hecht SS (1983) Cancer Res 43:3132
78. Salmeen IT, Pero AM, Zator R, Schuetzle D, Riley TL (1983) AMES-ASSAY – chromatograms, the identification of mutagens in diesel particle extracts. Ford Motor Company Research, P.O. Box 2053, Dearborn, Michigan
79. Ames BN, McCann J, Yamasaki E (1975) Mutat Res 31:347
80. DFG (Deutsche Forschungsgemeinschaft): Gesundheitsschädliche Arbeitsstoffe – Toxikologisch arbeitsmedizinische Begründung von MAK-Werten; edited by Henschler D. VCH Verlagsgesellschaft, Weinheim, Loseblattsammlung
81. Chou MW, Heflich RH, Casciano DA, Miller DW, Freeman JP, Evans FE, Fu PP (1984) J Med Chem 27:1156
82. Möller M, Thorgeirsson SS (1984) Proc Amer Assoc Cancer Res 25:110
83. Marshall TC, Royer RE, Li AP, Kusewitt DF, Brooks AL (1982) J Toxicol Environ Health 10:373
84. Andrews PA, Bryant D, Vitakunas S, Gouin M, Anderson G, McCarry BE, Quilliam MA, McCalla DR (1985) Metabolism of nitrated polycyclic aromatic hydrocarbons, formation of DNA-adducts in Salmonella thyphimurium. In: Cooke M, Dennis AJ (eds) Polynuclear Aromatic Hydrocarbons – Mechanism, Methods and Metabolism, p 89. Battelle Memorial Institute, Columbus, Ohio
85. Djuric Z, Fifer EK, Yamazoe Y, Beland FA (1988) Carcinogenesis 6:941
86. Wegenke M, Wanders H, Roscher E, Wolff T (1988) Formation of DNA-adducts in various cell lines after exposure to 1,3- and 1,6-dinitropyrene. Poster at Eurotox-Congress, München
87. Roy AK, El-Bayoumy K, Hecht SS (1989) Carcinogenesis 10:195
88. Tee LBG, Minchin F, Ilett KF (1988) Carcinogenesis 9:1869
89. Löfroth G (1981) Environ Science Res 22:319
90. Ohgaki H, Matsukura N, Morino K, Kawachi T, Sugimura T, Morita K, Tokiwa H, Hirota T (1982) Cancer Lett 15:1
91. Tokiwa H, Otofuji T, Nakagawa R, Horikawa K, Maeda T, Sano N, Izumi K, Otsuka H (1986) Dinitro derivatives of pyrene, fluoranthene in diesel emission particulates, their tumorigenicity in mice, rats. In: Ishinishi N, Koizumi A, McClellan RO, Stöber W (eds) Carcinogenic and mutagenic effects of diesel engine exhaust, p 253, Elsevier, Amsterdam
92. El-Bayoumy K, Hecht SS, Sackl T, Stoner GD (1984) Carcinogenesis 5:1449
93. Sato S, Ohgaki H, Takayama S, Ochiai M, Tahira T, Ishizaka Y, Nagao M, Sugimura T (1986) Carcinogenicity of dinitropyrenes in rats and hamsters. In: Ishinishi N, Koizumi A, McClellan RO, Stöber W (eds) Carcinogenic and mutagenic effects of diesel engine exhaust, p 272, Elsevier, Amsterdam
94. Wislocki PG, Bagan ES, Lu AYH, Dooley KL, Fu PF, Han-Hsu H, Beland FA, Kadlubar FF (1986) Carcinogenesis 7:1317
95. King LC, Loud K, Tejada SB, Kohan MJ, Lewtas J (1983) Environ Mutagen 5:577
96. McClellan RO (1986) Opening remarks: Toxicological effects of emissions from diesel engines. In: Ishinishi N, Koizumi A, McClellan RO, Stöber W (eds) Carcinogenic, mutagenic effects of diesel engine exhaust, p 3, Elsevier, Amsterdam
97. Pitts JN Jr, Katzenstein YA (1981) Atmospheric Environ 15:1782
98. Ioki Y (1977) J chem Soc Perkin Trans II:1240
99. Pitts JN Jr, van Cauwenberghe KA, Grosjean D, Schmid JP, Fitz DR, Belser WL Jr, Knudson GB, Hynds PM (1978) Science 202:515
100. Chapman OL, Heckert DC, Reasoner JW, Thackaberry SP (1966) J Am Chem Soc 88:5550
101. Grimmer G, Hildebrandt A, Böhnke H (1972) Probenahme und Analytik polycyclischer aromatischer Kohlenwasserstoffe in Kraftfahrzeugabgasen. Erdöl und Kohle – Erdgas – Petrochemie 25:442
102. Kraft J, Lies K-H (1981) Polycyclic aromatic hydrocarbons in the exhaust of gasoline and diesel vehicle. SAE-Paper No. 810082
103. Hartung A, Kraft J, Lies K-H, Schulze J (1982) Messung polycyclischer aromatischer Kohlenwasserstoffe im Abgas von Dieselmotoren. MTZ 43: S. 263

104. Lies K-H, Hartung A, Postulka A, Gring H, Schulze J (1986) Composition of diesel exhaust with particular reference to particle bound organics including formation of artifacts. In: Ishinishi N, Koizumi A, McClellan RO, Stöber W (eds) Carcinogenic and mutagenic effects of diesel engine exhaust, p 65, Elsevier, Amsterdam
105. VDI 3872 Page 1, Draft of a Guideline (1989a) Emission Management: Measurement of polycyclic aromatic hydrocarbons (PAH). Measurement of PAH in the Exhaust Gas from Gasoline, Diesel Engines of Passenger Cars – Gas Chromatographic Determination. Available from Beuth Verlag, D-1000 Berlin/F.R.G.
106. VDI 3973 Page 1, Draft of a Guideline (1989b) Emission measurement: Measurement of polycyclic aromatic hydrocarbons (PAH) in stationary industrial plants – Dilution method (RWTÜV method) – Gaschromatographic determination. Available from Beuth Verlag, D-1000 Berlin/F.R.G.
107. Lies K-H, Postulka A, Gring H (1984) Characterization of exhaust emissions from diesel-powered passenger cars with particular reference to unregulated components. SAE-Paper, No. 840361
108. Hayano S, Jang-Ho L, Furuya K, Kikuchi T, Someya T, Oikawa Ch, Iida Y, Matsushita H, Kinouchi T, Manabe Y, Ohnishi Y (1985) Atmospheric Environ 19:1009
109. Schuetzle D (1983) Environ Health Perspect 47:65
110. Gibson TL, Ricci AI, Williams RL (1981). Measurement of PAH, their derivatives and their reactivity in diesel automobile exhaust; p 707. In: Bjoerseth A, Dennis AJ (eds) Polynuclear aromatic hydrocarbons. Battelle Press, Columbus, Ohio
111. Hartung A, Kraft J, Schulze J, Kieß H, Lies K-H (1984) Chromatographia 19:269
112. Nielsen T, Ramdahl T, Bjoerseth A (1983) Environ Health Perspect 47:103
113. Brenner KS (1986) Chemosphere 15:1917
114. Schuetzle D, Perez JM (1983) Factors influencing the emissions of nitrated polynuclear aromatic hydrocarbons from diesel engines. JAPCA 33:751
115. Kopczynski SL (1987) Intern J Environ Anal Chem 30:1
116. Jin Z, Rappaport SM (1983) Anal Chem 55:1778
117. Tejada SB, Zweidinger RB, Sigsby JE Jr (1986) Anal Chem 58:1827
118. Henderson TR, Sun JD, Royer RE, Clark ChR, Lee AP, Harvey TM, Hunt DF, Fulford JE, Lovette AM, Davidson WR (1983) Environ Sci Technol 17:443
119. Henderson TR, Royer RE, Clark ChR, Harvey TM, Hunt DF (1982) J Applied Toxicol 2:231
120. Schuetzle D, Riley TL, Prater TJ (1982) Anal Chem 54:265
121. Bodine RS, Ruehle PH, Roth RW, Bosch G, Bosch L, Opperman G, Saugier JH (1981) Synthesis of selected nitro polycyclic aromatic hydrocarbons. Midwest Research Institute, 425 Volker Boulevard, Kansas City, Missouri 64110, USA
122. Tomkins BA (1985) Chromatographic detectors used for the determination of nitrated polycyclic aromatic hydrocarbons. In: White CM (ed) Nitrated polycyclic aromatic hydrocarbons. Huethig, Heidelberg – Basel – New York, 87–120
123. Jager J (1978) J Chromatogr 152:575–578
124. Paputa-Peck MC, Marno RS, Schuetzle D, Riley TL, Hampton CV, Prater TJ, Skewes LM, Jensen TE, Ruehle PH, Bosch LC, Duncan WP (1983) Anal Chem 55:1946–1954
125. Tanner RL, Fajer R (1983) Intern J Environ Anal Chem 14:231–241
126. White CM, Robbat A Jr, Hoes RM (1984) Anal Chem 56:232–236
127. Yu WC (1983) In: Cooke M, Dennis AJ (eds) Polynuclear aromatic hydrocarbons: Formation, metabolism, measurement, Battelle Press, Columbus, OH, 1267–1277
128. Krull IS, Swartz M, Hilliard R, Xie K-H, Driscoll JN (1983) J Chromatogr 260:347–362
129. Tomkins BA, Brazell RS, Roth ME, Ostrum VH (1984) Anal Chem 56:781–786
130. Gibson TL, Ricci AI, Williams RL (1981) In: Cooke M, Dennis AJ (eds) Polynuclear aromatic hydrocarbons: Chemical analysis, biological fate, Columbus, OH, Battelle Press, 707–717
131. Levine SP, Skewes LM, Abrams LD, Palmer AG III (1982) In: Cooke M, Dennis AJ, Fisher GL (eds) Polynuclear aromatic hydrocarbons: Physical and biological chemistry, Columbus, OH, Battelle Press, 439–448
132. Rappaport SM, Jin ZL, Xu XB (1982) J Chromatogr 240:145–154
133. Jin Z, Rappaport SM (1983) Anal Chem 55:1778–1781
134. Krull IS, Ding X-D, Selavka C, Bratin K, Forcier G (1984) J Forensic Sciences 29:449–463
135. Umweltbundesamt (ed) (1979) Luftqualitätskriterien für ausgewählte polyzyklische aromatische Kohlenwasserstoffe. Bericht 1/79, E. Schmid Verlag, Berlin
136. VDI-Berichte 358 (1980) Luftverunreinigung durch polyzyklische aromatische Kohlenwasser-

stoffe – Erfassung und Bewertung. Kolloquium Hannover 1979, VDI-Verlag GmbH, Düsseldorf

137. Lewtas J, Williamson K (1986) A retrospective view of the value of short-term genetic bioassays in predicting the chronic effects of diesel soot. In: Ishinishi N, Koizumi A, McClellan RO, Stöber W (eds) Carcinogenic, mutagenic effects of diesel engine exhaust, p 119, Elsevier, Amsterdam

138. Lepperhoff G (1981) PAH Emissionen von Ottomotoren (Part I). Wissenschaft und Umwelt 2:72

139. Schuetzle D, Frazier JA (1986) Factor influencing the emission of vapor, particulate phase components from diesel engines. In: Ishinishi N, Koizumi A, McClellan RO, Stöber W (eds) Carcinogenic, mutagenic effects of diesel engine exhaust, p 41, Elsevier, Amsterdam

140. Risby TH, Lestz SS (1983) Environ Sci Technol 17:621

141. Walsh MP (1983) Die Vorteile der Verwendung von bleifreiem Benzin und Katalysatoren zur Kontrolle der Abgase von Fahrzeugen. 874 North Livingston Street, Arlington, Virginia 22205, USA, 703-241-1297

Chlorinated Ethanes

K.G. Mross and J.K. Konietzko
Institut für Arbeits- und Sozialmedizin, Obere Zahlbacher Str. 67,
Johannes-Gutenberg-Universität, D-6500 Mainz

The Handbook of Environmental Chemistry,
Volume 3 Part G, Ed. O. Hutzinger
© Springer-Verlag Berlin Heidelberg 1991

Summary

Some chlorinated ethanes are produced in large quantities e.g. monochloroethane, 1,2-dichloroethane and 1,1,1-trichloroethane. The environmental fate and toxic effects of chlorinated ethanes vary substantially depending on their toxicokinetics. A human risk seems to exist only at the workplace, especially with 1,2-dichloroethane, 1,1,1- and 1,1,2-trichloroethane.

Environmental hazards are likely with compounds produced in large amounts although measurable concentrations have only been detected with 1,2-dichloroethane and 1,1,1-trichloroethane. A global environmental hazard with chlorinated ethanes can be excluded.

General Remarks

Chlorinated ethanes belong to the halogenated alkanes* and include the following substances:
– Monochloroethane,
– 1,1-Dichloroethane,
– 1,2-Dichloroethane,
– 1,1,1-Trichloroethane,
– 1,1,2-Trichloroethane,
– 1,1,1,2-Tetrachloroethane,
– 1,1,2,2-Tetrachloroethane,
– Pentachloroethane, and
– Hexachloroethane.

All chlorinated ethanes are of human origin. Up to now, no natural occurrence of these chemicals has been observed.

Technology and Economics

History

Whereas some of the substances were prepared in the 15th and 16th century, most chlorinated ethanes were first synthesized during the 19th century. Large-scale production, however, began in the first half of the 20th century. Since exact information is important for the assessment of environmental hazards, available data will be included in each section.

Physical and Chemical Properties

Although chlorinated ethanes are of similar structure, they differ slightly in their physical and chemical properties, depending on the number and position of the chlorine atoms. Increasing numbers of chlorine atoms raise the boiling point and heat of vaporization per mole; increasing density, viscosity, and surface tension lower the heat of combustion, and decrease solubility and inflammability. Boiling points and densities of asymmetric compounds are lower than those of their symmetric isomers. All chlorinated ethanes are readily soluble in organic solvents and only slightly miscible in water; they form azeotropes with both.

Production

Until this method of production became too expensive, chlorinated ethanes were produced from acetylene. In the last thirty years production has been based

* See also: Pearson, G.R.: G_1 and G_2 Halocarbons, Vol. 3/B, P.?

on ethylene. In these new technologies an excess of hydrogen chloride is produced. The problem of environmental load can be satisfactorily solved from a technical point of view by appropriate production processes (oxychlorination and hydrochlorination) and by residue utilization and incineration plants.

Impurities and Stabilizers

A few chlorinated ethanes are sensitive to physical and chemical influences, i.e. they are oxidized by light and oxygen and form hydrogen chloride, phosgene, carbon oxides, and acetyl chloride. Chlorinated acetic acids formed in a wet environment are highly corrosive. Most chloroethanes, therefore, must be stabilized prior to technical application. The most important stabilizing agents are amines, phenols, alcohols, and terpenes. Special stabilization properties are required for each field of application. Since most producers have developed special stabilizing systems for their products, it is impossible to provide general data on the type and amount of stabilizing agents. The most common stabilizing agents are included in each section. The impurities of chlorinated ethanes vary considerably, depending on the compound, producer, and technical requirements. As a rule, the higher the degree of purity, the greater the stability of the chlorinated ethane [1].

Use

Approximately half of the chlorinated ethanes are of major industrial importance. For thirty to forty years the market for these substances increased because of their unique properties with a maximum in the late 1970s. Increasing environmental consciousness and especially the use of unleaded gasoline and the replacement by other substances will lead to a decrease in the use of most chlorinated ethanes in the near future. However, there will be a market for these substances for the next decades. At present, they are used almost exclusively as initial and intermediate products for chemical synthesis processes and as solvents for cleaning and extraction procedures [2].

Environment and Biology

During production, transportation, storage, use, and disposal, chlorinated ethanes might be released into the environment. Most escape as vapors in the air and to a lesser degree into wastewater. Human exposure occurs directly through air and water and indirectly through animals and plants via the food chain. Only for a few chlorinated ethanes are measurements of atmospheric concentrations available. These measurements will be cited later. Estimated on the basis of their reaction with hydroxyl radicals, the tropospheric half-life ranges from 18 to 380 days. This is considerably shorter than the half-life of carbon tetrachloride or

Freons (8×10^4 days). Under laboratory conditions phosgene and hydrogen chloride — as well as monochloroacetyl chloride, dichloroacetyl chloride, and trichloroacetyl chloride — can be detected by the photooxidation of chlorinated ethanes.

It is conceivable that these compounds are formed in the higher layers of the stratosphere and affect the ozone layer. Chlorinated ethanes are heavier than water and will be poorly absorbed by soil particles. Therefore they quickly penetrate the soil layer and accumulate in ground water. A proportion of them may evaporate from the water surface. Under experimental conditions at the 1 ppm level in agitated water, the evaporation half-life is about 30 min for all chlorinated ethanes. On the whole, the bioconcentration of chlorinated ethanes is low, but it increases as the chlorine content increases. It has been shown that bioaccumulation is related to the octanol/water partition coefficient of the compound. Log(octanol/water partition coefficient) and other measurement data on the bioconcentration and biomagnification in environmental systems, so far available are included in the following. All the reported information tends to indicate that, in contrast to other halogenated hydrocarbons such as pesticides and Freons, chlorinated ethanes do not constitute a global threat to the environment. The potential threat is estimable and controllable. Local human hazards have been demonstrated at production and processing sites, particularly in workplaces. Here chlorinated ethanes are inhaled primarily as vapors. Percutaneous absorption is known, but is less important. Ingestion can occur either accidentally or intentionally. However, the chlorinated ethanes themselves are not the only hazard. Decomposition products are as dangerous as the original substances, sometimes more so (See dichloroacetylene or phosgene). In the presence of hot metal and open flames, e.g. welding equipment and lighted cigarettes, every chlorinated ethane is capable of forming phosgene. Higher concentrations of phosgene induce pulmonary edema with fatal results. Increased misuse of chlorinated ethanes by solvent "sniffing" has been reported. The number of undetected cases is estimated to be high. Exact statistics, however, are not available.

Pharmacokinetics

Chlorinated ethanes are lipophilic substances. The usual mode of human intake is by vapor inhalation. Skin absorption or ingestion by drinking water or eating is less important. The inhaled dose depends on the chlorinated ethane concentration in respiratory air and the duration of exposure. Additional factors such as respiratory minute volume, can modify this process. Given constant exposure, vapor uptake continues until concentration in external air, respiratory air, and body organs has reached a steady state. The tissue distribution rate in the human body depends on organ perfusion and solubility of the compound.

Regarding the perfusion rate, three distribution compartments have been described:

1. internal organs (good perfusion),
2. muscle tissue (moderate perfusion), and
3. fatty and connective tissue (poor perfusion).

Within these compartments concentrations are primarily governed by partition coefficients between individual organs. The blood/air partition coefficients increase in direct relation to the carbon chain length: they are markedly higher for isomers with a high boiling point (1,2-dichloroethane, 1,1,2-trichloroethane, 1,1,2,2-tetrachloroethane). The oil/blood partition coefficients react similarly. Distribution of chlorinated ethanes with high partition coefficients therefore is faster, and exhalation slower, than in chlorinated ethanes with lower partition coefficients.

Exhalation, however is not the only route of elimination. Parallel to this unaltered release, biochemical mechanisms of elimination are also operative. By these biochemical mechanisms the structure of chlorinated ethanes is altered, so that these ethanes are capable of coupling with endogenous substances of the intermediate metabolism and thereby rendered water soluble and excretable. Metabolism occurs primarily in the liver. It does not lead to detoxification of every compound; relatively harmless initial substances can also be activated to toxic metabolites. Compared with some of the aromatic chlorinated hydrocarbons used worldwide as pesticides, the biological half-life of chlorinated ethanes is short. Accumulation of higher concentrations in the human body or in experimental animals has not yet been detected.

Biomedical Effects and Toxicity

Toxic effects of chlorinated ethanes can be understood by knowing their pharmacokinetics. Human and nonhuman exposures show a distinction between nonspecific and specific effects; both are essentially dose dependent.

Nonspecific Effects

Nonspecific effects are uniform and homogenous reactions of the body, with no dependence on the chemical structure of the chlorinated ethanes, but they are dose dependent. These reactions include irritations of skin and mucosa, irritation of the central nervous system (CNS) or paralysis, and sensitization of the myocardial conduction system.

The skin, as a rule, is only reddened. Since chronic skin contact with liquid chlorinated ethanes is poorly tolerated, serious skin damage is usually avoided. Independent of this irritation, long-term exposure to vapor dries out the skin, making it susceptible to various allergenic substances in the occupational and private spheres. Direct allergization by chlorinated ethanes is rare.

Central nervous system effects are best described as a prolonged preanesthetic state which seldom progresses to an anesthetic state. Symptomatology is ambiguous, not characteristic, and often misleading. Depending on concentration and duration of exposure, affected individuals will complain of mild discomfort, loss of appetite, stomach pain, dizziness, fatigue, drunkeness, nausea and vomiting. Disorders of balance and motoric ataxia resembling alcohol intoxication at the advanced stage. Continuing uptake of chlorinated ethanes results in anesthesia and eventually fatal respiratory paralysis. Chlorinated ethanes are able to affect the central nervous system because of their excellent solubility in fat. The exact mechanism of action, however, is unknown. Analogous studies using chemically inert anesthetics tend to indicate, that the effect is produced primarily by physical changes at the synapses as well as by inhibition of synaptosome membrane-bound integral enzymes.

This means that the nonspecific effects on the central nervous system decrease in direct relation to the elimination; the effects subside completely within a few hours after exposure has been discontinued.

Specific Effects

Specific means that chlorinated ethanes preferentially damage certain organs with dependence on their chemical structure. Analogous to other chlorinated hydrocarbons, the specific organ effect is mainly a consequence of metabolism. Type and extent of metabolic transformation and metabolite determine the occurrence of organ damage and which organ will be affected. This explains, for example, the varying degrees of hepatotoxicity and nephrotoxicity.

However, all chlorinated ethanes are able to sensitize the myocardial conduction system to sympathetic impulses and sympathomimetics, the consequence being cardiac dysrhythmia. Dose dependent single or multiple extrasystoles have been observed; rare cases of fatal ventricular fibrillation have even been reported. The pathogenesis of this sensitization process is still unknown.

Mutagenicity, Teratogenicity, and Carcinogenicity

Estimates of cancer risk may be inferred from epidemiological studies and case reports or animal experiments. These studies are, however, elaborate and expensive, and therefore only a few such studies have been carried out with chloroethanes up to now. Short-term tests have been frequently used as screening procedures. These are based on the fact that most organic carcinogens initially affect DNA, i.e. the genetic material. There is no correlation between carcinogenicity on one hand, and teratogenicity or embryotoxicity on the other hand.

Therefore, all the available information about mutagenicity, teratogenicity and carcinogenicity will be shown in different sections. For practical reasons embryotoxic findings will be included in the teratogenicity section.

Potential Environmental and Occupational Hazards

Chlorinated ethanes may lead to following categories of hazards:
1. Global environmental hazards,
2. Local environmental hazards,
3. Hazards to limited groups of people.
Only for 1,1,1-trichloroethane a global environmental hazard seems possible because of its ability to degrade the ozone layer in the atmosphere.

Further investigation is necessary to clarify this hypothesis. Local environmental hazards are most likely with 1,1,1-trichloroethane and 1,2-dichloroethane. Both compounds are produced in major amounts and bring a measurable load to various ecological systems both near to and far from their production site. Concentrations detected up to now are in ranges that exclude any hazard to vegetation, animals, or humans, but our range of knowledge is based on only a few individual investigations in some countries. They should be enlarged by systematic measurements in air, water, and soil on an international basis.

Only a few results are available about the environmental load of monochloroethane. This fact is surprising as monochloroethane production is in second place among all chlorinated ethanes. Measurements of the atmospheric load seem particularly necessary, whereas the contamination of soil and water is probably very low. All other chlorinated ethanes are produced in smaller quantities which makes it unlikely that they are of any outstanding ecological importance.

Theoretically, hazards to limited groups of people may exist in all workplaces, where chlorinated ethanes are in use. NIOSH estimated that about 5 million workers are exposed to these substances. Because of their production rates only 1,1,1-trichloroethane, 1,2-dichloroethane, monochloroethane, and 1,1,2-trichloroethane are of major importance. All other chlorinated ethanes are only used in small quantities at workplaces. In individual cases they may cause severe intoxication. It is difficult to assess the general risk because of the large range of variations in exposure conditions, but workplace concentrations are usually in the part per million range whereas the exposure of general population is in part per trillion or part per billion range. A real human risk seems to be only in special workplaces and not for the general population.

Methods of Analysis and Permissible Exposure Standards

The most reliable methods for detecting chlorinated ethanes in air, water, and biological material are gas chromatography (GC) and gas chromatography/mass spectrometry (GC/MS). Because of its low sensitivity, infrared spectroscopy is suitable only at higher concentrations. The rapid development of analytical techniques makes it impossible to recommend any particular method for general use. Further information on analytical techniques can be seen in other sections.

There are several occupational exposure standards in the world. Most of them consider only the concentrations in the air and the absorption through the lungs and skin, i.e. normal conditions at the workplace. Because of individual differences in sensitivity, they are not sharp lines between harmless and dangerous concentrations. In most cases they are far below that what is known as the "No observed adverse effect level" (NOAEL).

In the United States of America there are some different levels to consider:

PEL (Permissible Exposure Limit) OSHA

OSHA Permissible Exposure Limits (PELs) are those limits, set by the Occupational Safety and Health Administration in the Department of Labor (OSHA), which have been at least described in the OSHA Final Rule Air Contaminants Permissible Exposure Limits [3].

REL (Recommended Exposure Limit) NIOSH

NIOSH develops and periodically revises recommendations or limits of exposure to potentially hazardous substances or conditions at the workplace. These recommendations are then published and transmitted to OSHA and the Mine Safety and Health Administration (MSHA) for use in promulgating legal standards [4].

The American Conference of Governmental Industrial Hygienists gave the first indices of allowable industrial exposure levels in 1938. This was for first time in the world and to now, these levels have been the most important occupational exposure levels (OELs) worldwide.

Threshold Limit Values (TLVs) and Biological Exposure Indices (BEIs) have been developed as guidelines to assist in the control of health hazards. These recommendations or guidelines are intended for use in the practice of industrial hygiene, to be interpreted and applied only by persons trained in this discipline. They are not developed for use as legal standards.

They are not for use, e.g. in the evaluation or control of community air pollution nuisances, in estimating the toxic potential of continuous, uninterrupted exposures or other extended work periods, as proof or disproof of an existing disease of physical condition, or adoption by countries whose working conditions differ from those in the USA. These limits set no fine lines between safe and dangerous concentrations nor are they a relative index of toxicity, and should not be used by anyone untrained in the discipline of industrial hygiene [5].

Three categories of Threshold Limit Values (TLVs) are specified:

TLV-TWA (Time weighted average) ACGIH
The Threshold Limit Value – Time Weighted Average (TLV-TWA) concentration for a normal 8-hour workday and a 40-hour work-week, to which nearly all workers may be repeatedly exposed, day after day, without adverse effects [5].

TLV-STEL (Short term exposure limit) ACGIH
The Threshold Limit Value – Short Term Exposure Limit (TLV-STEL) is the concentration, to which workers can be exposed continuously for a short period of time without suffering from irritation, chronic or irreversible tissue damages, or narcosis of sufficient degree to increase the likelihood of accidental injury, impair self-rescue or materially reduce work efficiently, and provided that the daily TLV-TWA is not exceeded. It is not a separate independent exposure limit, rather it supplements the time-weighted average limit (TWA) where there are recognized acute effects from a substance whose toxic effects are primarily of a chronic nature. STELs are recommended only where toxic effects have been reported from high short-term exposures in either humans or animals [5].

Threshold Limit Value – Ceiling (TLV-C) ACGIH
The Threshold Limit Value – Ceiling (TLV-C) is the concentration that should not be exceeded during any part of the working exposure [5].
The German MAK value corresponds approximately to the TLV-TWA.

BEI (Biological Exposure Indices)
Another way of assessing worker's exposure to chemicals is the so called biological monitoring. Its determinant is called BEI (Biological Exposure Index). It can be determined by the chemical itself or its metabolites or a characteristic, reversible biochemical change, induced by the chemical. Measurements can be made in exhaled air, blood, urine, or other biological specimen, collected from the exposed worker at a specific time.
BEIs (Biological Exposure Indices) are reference values as guidelines for the evaluation of potential health hazards and apply to eight-hour exposures, five days a week. They should not be applied either directly or through a conversion factor, in the determination of safe levels for non-occupational exposure to air and water pollutants, or food contaminants. And they are not intended for use as a measure of adverse effects or for diagnosis of occupational illness.
At present, there are no adopted biological exposure indices for chlorinated ethanes, but ACGIH intends to establish one for 1,1,1-trichloroethane.
The German BAT-Value (Biologischer Arbeitsstoff Toleranz Wert) corresponds approximately to the BEI.
The United States Environmental Protection Agency (EPA) have published standards for the protection of both, aquatic and human life for many contaminants. These standards are chosen, to protect both, the health of the aquatic organisms and their suitability for human consumption. Because of the differing toxicity in both media, separate criteria are given for fresh water and salt water. The 24 hour average concentration and the maximum ceiling concentration (MCC), which cannnot be exceeded, are cited for both criteria. Limit values for the protection of human health are calculated by the possible intake through drinking water and eating fish.

Chlorinated Ethanes

Monochloroethane

$$
\begin{array}{c}
\text{H} \ \ \text{H} \\
| \ \ \ | \\
\text{Cl--C--C--H} \\
| \ \ \ | \\
\text{H} \ \ \text{H}
\end{array}
$$

Synonyms

Chloroethane, Chloroethyl, Ethyl Chloride, Hydrochloric Ether, Muriatic Ether.

History

Monochloroethane does not occur as a natural product. As the first synthesized chlorinated hydrocarbon, it was produced in 1440 by Valentine in reacting ethanol with hydrochloric acid. Largescale production started in the 1930s with the increasing need for tetraethyl lead in the automotive industry. The trend to unleaded gasoline in most countries has led to a decrease in production [2, 6].

Physical Properties

Monochloroethane is a colorless thin liquid of pleasant etheral odor and burning taste, is volatile at room temperature and burns with a smoky greenish flame. It is readily soluble in many organic solvents and highly flammable. Its physical properties are shown in Table 1 [2, 6].

Chemical Properties

Monochloroethane hydrolizes or oxidizes gradually at room temperature. At temperatures above 400 °C it dehydrochlorinates to ethylene. Phosgene and hydrogen chloride are formed as combustion products. As the most reactive chlorinated ethane, monochloroethane is used as the starting material for many chemical syntheses [2, 6].

Production, Storage, and Transportation

Monochloroethane is produced almost exclusively by free radical chlorination of ethane or hydrochlorination of ethylenes. Other production processes are of no practical importance. The technical degree of purity determined by gas chromatography is 99.9%. If prepared from industrial methylated spirit, it contains a small variable amount of methyl chloride.

Monochloroethane can react vigorously with oxidizing materials. Contact with chemically active metals such as sodium, potassium, calcium, powdered aluminium, zinc, and magnesium may cause fires and explosions.

Therefore it is stored and transported in compression-proof steel tanks or canisters with a reflective coating to protect it against sunlight. Stabilizing agents are not necessary [2, 6].

Use

In 1984 approximately 80% of the consumption were used for the production of tetraethyl lead as an antiknocking compound. 15% went to ethyl cellulose production and 5% for miscellaneous use, such as synthesis of dyes and chemicals, refrigerants, extracting agents in temperature-sensitive odoriferous substances, solvents in the plastics industry, adhesive agents, and inks, as well as refrigeration anesthetics [2, 6, 7].

Pharmacokinetics

Monochloroethane is rapidly absorbed via the lungs. Approximately 75% is bound to cell constituents in blood and 25% to plasma [8]. The Ostwald solubility coefficient (blood/air) is 2.5; the oil/blood partition coefficient is 960 [9]. The highest concentrations are found in perirenal fatty tissue and the lowest in cerebrospinal fluid. Brain concentration is twice that of blood. Unaltered monochloroethane is exhaled rapidly and completely. Slight metabolism to ethanol by dechlorination can be detected only at high anesthetic concentrations [10, 11].

Table 1. Physical properties of monochloroethane

Molecular weight:	64.52
Density:	924 kg/m^3 (0 °C)
Vapor density:	2.76 kg/m^3 (20 °C)
Vapor pressure:	403 mbar (− 10 °C)
	623 mbar (0 °C)
	929 mbar (10 °C)
	1,342 mbar (20 °C)
Melting point:	− 138 °C
Boiling point:	12 °C
Spontaneous ignition temperature:	517 °C
Explosion limit in air:	3.6–14.8 vol%
Solubility in water:	0.455 mass% (0 °C)
Solubility of water in Monochloroethane:	0.07 mass% (0 °C)
Log (octanol/water partition coefficient):	1.43
1 ppm = 2.681 mg/m^3	1 mg/m^3 = 0.373 ppm

Biomedical Effects and Toxicity

- Animal Toxicity

No effects have been observed in guinea pigs after 13 h inhalation of 10,000 ppm; however, abnormal behaviour, histological alterations in liver and kidneys, and, in the presence of an open flame, severe pulmonary irritations have been observed at concentrations above 20,000 ppm [12, 13, 14]. The narcotic concentration in cold-blooded animals was approximately 18,000 ppm, in rodents it started at 36,000 ppm [15, 16]. Concentrations above 100,000 ppm led to complete anesthesia in all animals [10, 13, 14]. Conclusive animal tests on chronic toxicity are not available.

- Human Toxicity

While acute irritations of skin and mucosa are slight, narcotic effects are pronounced. Mild complaints from subjects may occur after 20 min with exposures above 13,000 ppm [17]. Inebriation and impaired coordination have been observed at concentrations above 20,000 ppm and 25,000, respectively. The stage of anesthesia is noticeable at 36,000 ppm [14, 17]. Respiratory arrest has been observed at 60,000 ppm [18]. Isolated deaths have been reported after short anesthesia. Since monochloroethane is absorbed and eliminated rapidly, it was formerly used as a short-time anesthetic. It has not been used as a general anesthetic, however, for more than 50 years because of its poor controllability, its low narcotic range, and the danger of overdosage. The presented observations were published 30 to 40 years ago; current information is not available.

Since monochloroethane is not metabolized, specific organ effects are highly unlikely. Only one case of reversible cerebellar dysfunction in a 28-year old woman who used monochloroethane as a narcotic for several months has been reported [19]. Occupational intoxication following chronic inhalation has not been described and is highly unlikely, since the fire hazard and low explosion limits eliminate the possibility of higher workplace concentrations. Epidemiological studies on larger samples are not available.

- Mutagenicity, Teratogenicity, and Carcinogenicity

There are no studies available on the mutagenicity, carcinogenicity and teratogenicity of monochloroethane.

Sampling and Analysis

Monochloroethane can be analyzed using gas chromatography with hydrogen-air flame ionization detection [22].

Environmental Fate

- General Environment

Production and utilisation of monochloroethane may result in environmental contamination. When released to the atmosphere, the predominant environmental process will be reaction with photochemically generated hydroxyl radicals with a half-life of about 40 days [20].

The tropospheric half-life is estimated to be 18 days [21]. If released to surface water, volatilization will be the dominant process as half-lives ranging from 1.1–5.6 days have been predicted for representative bodies of water. In ground water, where volatilization does not occur, hydrolysis may be the most important removal mechanism. The hydrolysis half-life has been estimated to be 38 days at 25 °C. Very limited biodegradation data suggest that monochloroethane may be biodegradable. But insufficient data are available to estimate the relative importance of biodegradation in the environment.

Aquatic bioconcentration, adsorption, direct photolysis, and oxidation are not important. If released to soil, monochloroethane will evaporate rapidly, where release to the air is possible [22].

- Workplace Environment

Production, storage, transportation, and use of monochloroethane may be hazardous. The United States National Institute for Occupational Safety and Health (NIOSH) estimates that approximately 113,000 workers in the United States of America are exposed to monochloroethane [23]. The risk in production, storage, and transportation, however, is slight. The free use of high doses, especially as a local refrigeration anesthetic in medical operations is more dangerous. Recent studies indicate an increase in the use of monochloroethane as a "sniffing" drug [19].

Table 2. Permissible exposure standards of monochloroethane

Index	Permissible exposure ppm (mg/m³)	Reference
TLV-TWA (ACGIH)	1,000 (2,600)	[5]
TLV-STEL (ACGIH)	not adopted	[5]
TLV-C (ACGIH)	not adopted	[5]
BEI (ACGIH)	not adopted	[5]
PEL (OSHA)	not established	[3]
(Final rule limits: TWA: 1,000 (2,600)		
STEL: – Ceil: –		
REL (NIOSH)	not recommended	[4]
MAK (GERMANY)	1,000 (2,600)	[24]
EPA criteria:	not established	[25]

1,1-Dichloroethane

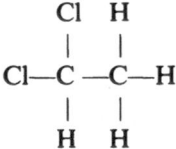

Synonyms

α-Dichloroethane, ethylidene chloride, ethylene dichloride, chlorinated hydrochloric ether.

History

1,1-Dichloroethane does not occur naturally. No information is available about the early history of its synthesis and production.

Physical Properties

1,1-Dichloroethane is an oily colorless liquid of sweet chloroform-like odor, and a saccharin-like taste. It is soluble in all liquid chlorinated hydrocarbons and in most organic solvents [2, 6, 7, 26].

Chemical Properties

Under dry conditions and at ambient temperatures 1,1-dichloroethane is extremely stable. At higher temperatures, hydrogen chloride is eliminated [2, 6]. Above 150 °C and in the presence of catalytic amounts of chlorine and iron, it dehydrochlorinates to vinyl chloride.

Production, Storage, and Transportation

As the less important substance of the two dichloroethane isomers, it often occurs as an unwanted byproduct in many chlorination and hydrochlorination processes of C_2 hydrocarbons. The most important industrial production process is as an intermediate in the production of 1,1,1-trichloroethane [2, 6].

Use

The most important role of 1,1-dichloroethane is its use as an intermediate in the production of 1,1,1-trichloroethane. Other possible occurrences is that as

Table 3. Physical properties of 1,1-dichloroethane

Molecular weight:	98.97
Density:	1,176 kg/m^3
Vapor pressure:	93 mbar (0 °C)
	153 mbar (10 °C)
	242 mbar (20 °C)
	369 mbar (30 °C)
Melting point:	− 97 °C
Boiling point:	57 °C
Explosion limit in air:	5.9–15.9 vol%
Solubility in water:	0.5 mass%
Solubility of water in 1,1-dichloroethane:	0.009 mass% (20 °C)
Log (octanol/water partition coefficient):	1.79
1 ppm = 4.113 mg/m^3	1 mg/m^3 = 0.243 ppm

a solvent for plastics, oils, and fats, as a cleaning agent, degreaser, as a fumigant and insecticide spray in rubber cementing [2, 6, 7, 23].

Pharmacokinetics

There is little information about uptake, distribution, metabolism and elimination of 1,1-dichloroethane and its possible metabolites. Formation of acetic acid has been speculated [11]. Blood/air and oil/blood partition coefficients are 4.7 and 40, respectively [27].

Biomedical Effects and Toxicity

− Animal Toxicity

In mice, the acute narcotic concentration is 8,000 to 10,000 ppm; the lethal concentration is 17,300 ppm [28, 29, 30]. The acute oral LD$_{50}$ is 14.1 g/kg body weight [31]. In two animal experiments, specific organ effects have been studied:

Whereas rats, guinea pigs, rabbits, and cats showed no clinical signs of damage after 3 months inhalation of 500 ppm, kidney damage was demonstrable in cats after a subsequent increase to 1,000 ppm. No pathologic alterations were demonstrated in rats, guinea pigs, rabbits, and dogs, exposed to 500 and 1,000 ppm for more than 6 months. Kidney and liver damage, however, were observed after a considerable increase in dose [32].

− Human Toxicity

At the beginning of this century 1,1-dichloroethane was used as an anesthetic. Information about concentrations and complications during anesthesia is not available. 1,1-dichloroethane-induced occupational diseases are not reported.

– *Mutagenicity*

In a BALB/c-3T3 cell transformation assay 1,1-dichloroethane was negative when conducted in the absence of an exogenous metabolic activating system [22].

– *Carcinogenicity*

Animal experiments on carcinogenicity have yielded negative results. However, due to the high death rate of the laboratory animals, a definite evaluation will not be possible until more studies are available [33].

– *Teratogenicity*

After exposure of pregnant Sprague-Dawley rats to subnarcotic concentrations for 7 h a day during days 6–15 of gestation no embryotoxic or fetotoxic effects were observed [34].

Sampling and Analysis

1,1-Dichloroethane was determined using gas chromatography-headspace analysis and GC/MS [22].

Environmental Fate

– *General Environment*

During production, further utilization, and disposal of 1,1-dichloroethane, environmental hazards are possible. However, measurements of its levels in air, water, soil, and food are not available. Its half-life in the troposphere is estimated to be 31 days [20].

Table 4. Permissible exposure standards of 1,1-dichloroethane

Index	Permissible exposure ppm (mg/m^3)	Reference
TLV-TWA (ACGIH)	200 (810)	[5]
TLV-STEL (ACGIH)	250 (1,010)	[5]
TLV-C (ACGIH)	not adopted	[5]
BEI (ACGIH)	not adopted	[3]
PEL (OSHA)	100 (400)	[3]
(Final rule limit: TWA: 100 (400), STEL: – Ceil: –		
REL (NIOSH)	100 (400)	[4]
MAK (GERMANY)	100 (400)	[24]
EPA criteria:	not established	[25]

Its low production rates exclude the possibility of major environmental hazards.

– Workplace Environment

Production, storage, transportation and further utilization may result in health hazards. NIOSH estimates that approximately 4,600 workers are exposed to 1,1-dichloroethane in the U.S.A. [23].

1,2-Dichloroethane

```
    H  Cl
    |   |
  H–C–C–H
    |   |
    Cl  H
```

Synonyms

Ethylene chloride, ethylene dichloride, sym-dichloroethane, α,β-dichloroethane, glycol dichloride.

History

1,2-Dichloroethane does not occur naturally. The first synthesis was achieved in 1795. Large-scale production started in the 1960s [2, 6, 35].

Physical Properties

1,2-Dichloroethane is a clear colorless, oily liquid of pleasant chloroformlike odor and sweet taste. It is slightly flammable and readily soluble in most organic solvents [2, 6, 36].

Chemical Properties

In the dry state, pure 1,2-dichloroethane is stable at higher temperatures. At temperatures above 600 °C it decomposes to vinyl chloride, and acetylene. Presence of light, air, and moisture lead to decomposition and the release of hydrogen chloride. Photooxidation may result in the formation of phosgene, carbon monoxide, and hydrogen chloride [2, 6, 22].

Production, Storage, and Transportation

At present 1,2-dichloroethane is a chemical with one of the highest production rates. During the past 20 years, there have been annual growth rates of more than 10%.

1,2-dichloroethane is industrially produced by chlorination of ethylene. This can either be carried out by using chlorine (direct chlorination) or hydrogen chloride (oxychlorination) as a chlorinating agent. The product by direct chlorination is nearly free of impurities, while during oxychlorination there are several impurities such as nitrogen dioxide, carbon monoxide, carbon dioxide, small amounts of ethylene, and 1,1-dichloroethane. The environmental load of 1,2-dichloroethane by the oxychlorination procedure is twice as high as in direct chlorination [2, 6, 22, 37]. 1,2-dichloroethane usually has the least impurities among chlorinated ethanes and contains only traces of 1,1,2-trichloroethane. At a water concentration of 50 ppm it is so stable, that transport without stabilizing agents is possible. As a rule, however, commercial 1,2-dichloroethane is stabilized with 0.1% alkylamines. It can be stored and transported in iron and stainless steel tanks, but not in aluminium drums [2, 6].

Use

At least 85% of 1,2-dichloroethane production is used for the production of vinyl chloride, 10% for the production of chlorinated solvents such as 1,1,1-trichloroethane and tri- and tetrachloroethylene. The rest is needed for products, like ethylendiamines. Only small amounts are used in dewaxing, deparaffinizing, petroleum fractions, and coating remover. Its use as a lead scavenger in leaded gasoline declines, while unleaded gasoline is increasing [2, 6, 22, 37, 38].

Table 5. Physical properties of 1,2-dichloroethane

Molecular weight:	98.97
Density:	1,281 kg/m^3
Vapor pressure:	33 mbar (0 °C)
	85 mbar (20 °C)
	320 mbar (50 °C)
	930 mbar (80 °C)
Melting point:	− 35 °C
Boiling point:	83 °C (1 bar)
Spontaneous ignition temperature:	440 °C
Explosion limit in air:	6.2 – 16.0 vol% (20 °C, 1.013 bar)
Solubility in water:	0.873 mass% (0 °C)
Solubility of water in 1,2-dichloroethane:	0.16 mass% (20 °C)
Log (octanol/water partition coefficient):	1.48
1 ppm = 4.11 mg/m^3	1 mg/m^3 = 0.243 ppm

Pharmacokinetics

There is no information about uptake, distribution, metabolism, and elimination of 1,2-dichloroethane in man.

Partition coefficients are: blood/air 19.5, and oil/blood, 23 [27]. 1,2-Dichloroethane is readily inhaled or ingested. To a lesser extent it is absorbed through the skin [44]. In inhalation and feeding tests with rats, equilibrium was established within 2–3 h, i.e., relatively slowly. After exposure was terminated, the blood concentration rapidly declined in a biphasic elimination curve. Half-lifes for the two phases were 6 and 35 min. Most of the 1,2-dichloroethane had been eliminated 18 h after exposure was terminated. Following inhalation or ingestion, labelled 1,2-dichloroethane is distributed primarily in the liver and kidneys. In the first 48 h after exposure, 30% of ingested, but only 1.8% of inhaled 1,2-dichloroethane was exhaled unaltered. Most of the 1,2-dichloroethane was metabolized and excreted in urine: 7.7% was exhaled as carbon dioxide. Within 48 h after inhalation, 96% of the active substance had been eliminated [39].

The findings after intravenous and intraperitoneal injection were similar. Major metabolites are chloroacetic acid, S-carboxymethylcystein, and thiodiacetic acid [40]. Another metabolite, chloroethanol, was detected in blood and liver of rats [22]. Decomposition to ethylene has been demonstrated [41]. Cytochrome P-450 and UDP-glucuronyltransferase activity in rat liver microsomes decreased markedly under the influence of 1,2-dichloroethane [42].

Rat liver microsomes are able to transform 1,2-dichloroethane into reactive metabolites that form covalent bonds with macromolecules, particularely DNA, RNA, and proteins. Studies with labelled 1,2-dichloroethane indicated that these covalent bindings were not limited to organs predisposed to cancer; they were also demonstrated in many other kinds of tissue. Pretreatment with phenobarbitone intensifies this binding [22, 39, 43].

Biomedical Effects and Toxicity

– Animal Toxicity

The 30 min exposure LD_{50} in rats was 12,000 ppm, 7 h exposure LD_{50} only 1,000 ppm. None of the rats survived 20,000 ppm [45]. Short anesthesia concentrations are attained with 5,000 ppm [28]. Anesthesia with 1,2-dichloroethane is deeper and more prolonged than with carbon tetrachloride or chloroform. Long-term exposure to concentrations of 500 ppm was lethal for rabbits, guinea pigs, and rats, but not for dogs, cats, and monkeys [32, 46].

Many different kinds of damage (i.e., liver and kidney damage, pulmonary edema, adrenal necrosis, intestinal and mesenteric hemorrhages, extremity paralysis, and myocardial necrosis) were observed in individual experimental animals after short- and long-term tests. Impairment of liver function was usually less severe and regressed more rapidly than after comparable carbon tetrachloride poisoning [14, 45, 46, 47]. Nubeculae could be induced in many experimental animals by inhalation exposure and subcutaneous injection; nubeculae regressed within a week to several months. These alterations were observed in dogs after inhalation exposure as low as 1,000 ppm and in foxes after inhalation exposure to 3,000 ppm. Since nubeculae also developed when the eyes were covered during inhalation, but did not form after direct instillation in the lids, these lesions must be systemic [47, 48, 49].

– Human Toxicity

1,2-Dichloroethane irritates the skin and mucosa. Severe necrotic alterations were regularly found in the gastrointestinal tract after accidental or intentional peroral intoxication [50, 51, 52, 53, 54]. Inflammatory and necrotizing alterations were detected along the respiratory tract after inhalation intoxication. The critical anesthesia limit has not been established, but it is probably around 20,000 ppm [55, 56]. Specific organ alterations develop after acute intoxication and chronic exposure, particularly after accidental or intentional intake of liniments containing 1,2-dichloroethane. Degenerative alterations were detectable in the liver and kidneys of a child who died 21 h after intake [35]. Liver and kidney damage were also found in most other cases of fatal poisoning. Degenerative alterations have been observed in the brain and myocardium. The frequency of cutaneous bleeding and organ hemorrhages (consumption coagulopathy) was remarkable. Liver damage was frequently observed in survivors of acute intoxication [42, 45, 52, 53, 57, 58, 59, 60, 61]. In one case, an irreversible cerebral defect with myoclonic syndrome, epileptic seizures, and permanent mental defects developed after peroral intoxication [62]. Irreversible organ damage after chronic exposure has not been reported. Nubeculae have not been observed in man.

– Mutagenicity

Mutagenicity tests with S. Typhimurium TA 1530, TA 1535, and TA 100 established 1,2-dichloroethane to be a weak mutagen without activation. The mutagenic effect was markedly intensified by the addition of cytosol and glutathione. Mutagenic effects were demonstrated in Escherichia coli, but not in Streptomyces coelicolor and Aspergillus nidulans [63, 64, 65, 66, 67, 68]. Increased mutation rates were observed in barley [69]. Sex-linked recessive lethals were found in Drosophila melanogaster. Chloroacetaldehyde, a presumed metabolite of 1,2-dichloroethane, proved to be a potent mutagen in S. typhimurium TA 100 [65].

– Carcinogenicity

There are no casuistic or epidemiological studies of the carcinogenicity of 1,2-dichloroethane in man. A few animal experiments, however, have been carried out. After feeding 1,2-dichloroethane to mice, the incidence of benign and malignant tumors was significantly higher in the treated group than in the control group and it was dose dependent. Results in rats were more diverse: Male animals developed squamous cell carcinomas of the forestomach and hemangiosarcomas more frequently than the female animals; the incidence of mammary adenocarcinoma and fibroadenoma, however, was higher in the female animals. The effect in mice after intraperitoneal administration was not clear [33, 70, 71].

A 2-year inhalation study using Swiss mice and Sprague-Dawley rats showed no increase in tumor incidence after exposure to concentrations of 5, 10, 50, and 150 ppm [72].

– Teratogenicity

1,2-Dichloroethane passes the placental barrier and accumulates in fetal tissue [73]. Inhalation tests in rats showed a 27.9% higher embryonic death rate in the exposed group than in the control group [74]. A further study conducted over a 6-month period revealed reduced fertility and increased perinatal mortality in the first generation [73], but not in the second one. Neither teratogenic nor embryotoxic damage, however, was observed after long-term exposure of pregnant rats and rabbits to 100 or 300 ppm of 1,2-dichloroethane [75].

Sampling and Analysis

Air samples were trapped with activated charcoal [76], desorbed, and analyzed with GC/FID [77, 78]. Drinking water was analyzed directly with GC/ECD [162]. Waste water and food required pretreatment [80, 81].

Environmental Fate

– General Environment

Production, storage, transport, and disposal of 1,2-dichloroethane may result in environmental hazards. The half-life of 1,2-dichloroethane in the atmosphere is estimated to be several weeks to months [82]. Attempts to detect 1,2-dichloroethane in the atmosphere have been unsuccessful. Concentrations in the vicinity of production plants, however, have been between 0.1 and 186 ppb. The greater the distance between measuring points and plants, the lower the values, i.e., within the parts per trillion (ppt) range [83, 84]. On the basis of measurement results and the population density in the vicinity of production plants, it has been estimated, that approximately 12.5 million people are exposed to annual mean 1,2-dichloroethane concentrations of 0.01 to 10 ppb; only 300,000, however, are exposed to concentrations exceeding 1 ppb [85, 86]. Concentrations of 0.01 to 9.4 ppb were obtained in major Japanese cities [87].

Waste water and waste are contaminated by 1,2-dichloroethane-bearing tar. These complex tar compounds are composed predominantly of aliphatic halogenated hydrocarbons, which may include 33% 1,2-dichloroethane or 1,1,2-trichloroethane, and 0.06% vinyl chloride [88]. Because of its high vapor pressure, 1,2-dichloroethane, insofar as it is not firmly bound in tars, volatilizes. The remainder is highly stable and slightly soluble in water; 1 part 1,2-dichloroethane dissolves in approximately 120 parts of water. Degradation is probably gradual. Exact data on the half-life, however, are not available. This gradual degradation is probably the reason for detection of 1,2-dichloroethanes in industrial waste waters as well as in river and tap water. An analysis of drinking water in 80 United States cities revealed concentrations as high as 6 μg/liter in 28 cases, and even 8 mg/liter in one case [89]. In another study, random samples taken from surface water in the vicinity of industrial plants revealed concentrations of 1 to 2 ppb in 53 of 204 samples [90]. Measurement values for river and tap water in Europe and Japan were below 1 μg/liter [91]. It is not entirely impossible that part of this 1,2-dichloroethane concentration was produced secondarily through the chlorination of drinking water. Activated charcoal filters nevertheless trap 90 to 100% of the 1,2-dichloroethane.

High bioaccumulation in aquatic plants or animals is improbable, even though 1,2-dichloroethane has occasionally been detected in fish and oysters. 1,2-dichloroethane, however, was also identified in industrial waste, dumped in the North Sea [88].

A slight risk is associated with the disinfection of food. Several weeks after disinfection, the 1,2-dichloroethane concentration in sacked wheat ranged from 23 to 43 ppm [9]. Concentrations of 2 to 23 μg/g were found in several spices that were extracted with 1,2-dichloroethane [92].

A high level of food contamination could not be demonstrated, but there is an other way of possible environmental distribution: in 1977, registration of pesticides containing 1,2-dichloroethane in the United States numbered 84. In total, these pesticides contained more than 2 million pounds of 1,2-dichloroethane [37].

– Workplace Environment

Workplace hazards may occur in the production and further utilization of 1,2-dichloroethane. Because of its wide range of use, the number of workers exposed to this substance is difficult to assess. NIOSH estimated an exposure of two million workers to 1,2-dichloroethane in at least 60 branches of industry, of whom 18,000 or more are exposed to high concentrations [23].

Since in these applications 1,2-dichloroethane is released into air, the most dangerous exposure is contact to cleaning agents and solvents in the metal and textile industry, and to disinfectants. Previously, antirheumatic ointments were a hazard for patients and medical staff. Antirheumatic ointments containing up to 80% 1,2-dichloroethane were available on the German and Swiss drug markets until just a few years ago.

Table 6. Permissible exposure standards of 1,2-dichloroethane

Index	Permissible exposure ppm (mg/m³)	Reference
TLV-TWA (ACGIH)	10 (40)	[5]
TLV-STEL (ACGIH)	not adopted	[5]
TLV-C (ACGIH)	not adopted (susp. carcinogen)	[5]
BEI (ACGIH)	not adopted	[5]
PEL (OSHA)	50 / 100 ceiling / 200 5 min 3 h peak	[3]
REL (NIOSH)	8 h time weighted 50 ppm ceiling 100 ppm max peak for 5 min in 3h of 8h shift 200 ppm	[4]
MAK (GERMANY)	20 (80)	[24]
EPA criteria:		[25]
freshwater aquatic life:	3,900 μg/liter (24 h average) 8,800 μg/liter (MCC)	
saltwater aquatic life:	880 μg/liter (24 h average) 2,000 μg/liter (MCC)	
human health:	7 μg/liter	

12.5 million people are estimated to be exposed to average annual concentrations of 0.009 to 9 ppb in the vicinity of production facilities. The exposure estimated from filling tanks with gasoline is 0.1 ng/day [25].

1,1,1-Trichloroethane

$$
\begin{array}{ccc}
Cl & & H \\
| & & | \\
Cl-C- & C-H \\
| & & | \\
Cl & & H
\end{array}
$$

Synonyms

α-Trichloroethane, chloroethene, methyl chloroform, methyl trichloromethane.

History

1,1,1-Trichloroethane does not occur naturally. First synthesized in 1840 by V. Regnault, it was not used industrially for more than 100 years. Frequently it was found as an unwanted byproduct. Commercial production started in 1950. After development of effective stabilizing systems, 1,1,1-trichloroethane became one of the major solvents for cold and vapor degreasing as well as several other applications [2, 6, 7].

Physical Properties

1,1,1-Trichloroethane is a colorless nonflammable liquid with a sweetish chloroformlike odor. It is miscible with all common organic solvents and is a good
solvent for grease, paraffin, wax and many other organic substances [2, 6, 7, 93].

Chemical Properties

1,1,1-Trichloroethane is highly unstable. It splits off hydrogen chloride at room
temperature and decomposes into 1,1-dichloroethane and hydrogen chloride at
400 °C. In the presence of metallic salts, this process starts at 150 °C. Temperature dependent oxidation to phosgene occurs in air. Hydrogen chloride, phosgene, and dichloroacetylene are produced under UV-radiation or elevated
temperature.
    1,1,1-Trichloroethane is extremely corrosive to aluminium. Inhibitors must
inevitably be used. Dry 1,1,1-trichloroethane moderately corrodes iron and zinc.
Corrosion, however, increases with the water content [2, 6, 7].

Production, Storage, and Transportation

Three different routes of industrial production are in use:
1. Thermal or photochemical chlorination of 1,1-dichloroethane,
2. Consecutive hydrochlorination of 1,1,2-trichloroethane via 1,1-dichloroethy-
   lene, and
3. Direct chlorination of ethane.
Several other processes have been proposed, but are not used on an industrial
scale.
    Since 1,1,1-trichloroethane is easily decomposed, it must be stabilized during
production. Different requirements for stabilization against metals, light, air,
and temperature have contributed to the development and patenting of many
stabilizing systems. These systems usually contain glycol diesters, ketones,
nitriles, dialkyl sulfoxides, dialkyl sulfides, dialkyl sulfites, tetraethyl lead, nitro-
aliphatic hydrocarbons, 2-methyl-3-butyl-2-on, tert-butyl alcohol, 1,4 dioxan,
dioxolane, sec-butyl alcohol, and monohydric acetylenic alcohols (approximately 3 to 7%). 1,1,1-Trichloroethane must be specially stabilized for metal
decreasing in the vapor phase [6]. A recently published study reported the
following impurities in 22 samples of technical 1,1,1-trichloroethane:

1,1-dichloroethylene	(30–900 µg/ml),
1,1-dichloroethane	(11 µg/ml),
trichloroethylene	(12 µg/ml),
1,1,2-trichloroethane	(9 µg/ml).
Further,	
Nitromethane	(18 µg/ml),

1,2-epoxybutane   (19 μg/ml),
butanol   (7 μg/ml), and
dioxane   (10 μg/ml)

have been identified as stabilizing agents [77].

Stabilized 1,1,1-trichloroethane can be stored in iron steel containers or, preferably in high-grade steel tanks to prevent contact with light, air, and moisture. 1,1,1-Trichloroethane should be transported in phosphatized, galvanized, or baked enamel tanks of high grade steel.

European and U.S. production rates of 1,1,1-trichloroethane in 1984 were estimated to be about 450,000 tons. In Europe, it competes strongly with trichloroethylene and tetrachloroethylene, and may replace these solvents in several applications [2, 6].

Use

1,1,1-Trichloroethane is primarily used as an industrial solvent in numerous applications, e.g. cold and hot cleaning and vapor degreasing. Formulations are used as solvents for adhesives and metal cutting fluids.

New applications have been found in textile processing and finishing and in dry cleaning, where 1,1,1-trichloroethane is replacing the widely used tetrachloroethene.

Special grades are used for the development of photoresists in the production of printed circuit boards. Because of its lower toxicity, 1,1,1-trichloroethane has replaced trichloroethylene in many fields, especially in the United States. In Europe, however, trichloroethylene has maintained its leading position, sometimes tetrachloroethylene.

Further advantages of 1,1,1-trichloroethane are its high solubility which allows it to be used even in very sensitive areas, its good evaporation rate, and the fact, that there is no fire or flash point [2, 6, 7, 42, 94, 95].

Table 7. Physical properties of 1,1,1-trichloroethane

Molecular weight:	133.42
Density:	1,371 kg/m^3 (0 °C)
Vapor pressure:	16 mbar (− 20 °C)
	49 mbar (0 °C)
	133 mbar (20 °C)
	320 mbar (40 °C)
	627 mbar (60 °C)
Melting point:	− 33 °C
Boiling point:	74 °C (1 bar)
Explosion limit in air:	8 vol% (25 °C, 1 bar)
Solubility in water:	0.44 mass% (20 °C)
Solubility of water in 1,1,1-trichloroethane:	0.05 mass% (20 °C)
Log(octanol/water partition coefficient)	2.17
1 ppm = 5.56 mg/m^3	1 mg/m^3 = 0.24 ppm

Pharmacokinetics

1,1,1-Trichloroethane can be inhaled, ingested or absorbed through skin [96]. Partition coefficients are: blood/air 3.3, and oil/blood, 108 [27].

In mice, 1,1,1-trichloroethane was distributed in the organs (primarily to the liver, then, in approximately equal amounts to brain and kidneys) fairly soon after uptake [19]. Tests with labelled 1,1,1-trichloroethane in rats showed that 98.7% was exhaled unaltered within 25 h after exposure. 0.5% was found as carbon dioxide, and 0.85% appeared in the urine as trichloroethanol [97]. In contrast to similar compounds, in vitro dechlorination was not demonstrable in the presence of liver microsomes, NADPH, and oxygen [98]. Some researchers, however, found small amounts of trichloroethanol and trichloroacetic acid in laboratory animals after inhalation and intraperitoneal injection [99]. These metabolites are probably produced by hydroxylation.

Similar mechanisms of uptake, distribution, and elimination may be assumed in man. Percutaneous absorption has been quantitatively demonstrated in man; the amount, however, is not significant [100]. After short-term inhalation, labelled 1,1,1-trichloroethane is exhaled rapidly and almost entirely. In comparison to other chlorinated hydrocarbons, this relative rapid exhalation may be explained by the low blood/air partition coefficient of 5 [101]. Results of an 8 h inhalation test of 72 to 213 ppm showed, that 90% of the inhaled dose had been exhaled within 8 days. Within the next 12 days, 1.8% of the retained dose was excreted in the urine as trichloroacetic acid and 4.2% as trichloroethanol [102]. Similar findings were also obtained in printers, exposed to mean workplace concentrations of 4 to 53 ppm; concentration of metabolites increased markedly towards the end of the working week [103]. The biological half-life of 1,1,1-trichloroethane depends primarily on the duration and level of exposure and is estimated to be 4 to 26 h [103, 104, 105, 106]. The uptake can be tripled by physical exertion (increased respiratory minute volume) [107].

Biomedical Effects and Toxicity

– Animal Toxicity

In comparison to other chlorinated hydrocarbons, there is only a mild irritant effect on the skin and mucosa of rabbits [108]. After 4 days inhalation of 500 ppm, rats showed no sign of abnormal behavior. Prolonged inhalation exposure, however, led to a decrease of RNA content in brain and P-450, microsomal hepatocytochrome [109].

Light anesthesia is attained at concentrations as low as 5,000 ppm. All animals survived short-term anesthesia concentrations of 18,000 ppm [110]. Oral LD_{50} in rats and mice is 11 g/kg body weight; intraperitoneal LD_{50} in mice is 5 mg/kg body weight [111, 112].

Organ damage has been reported after acute as well as chronic inhalation. McNutt et al. exposed mice to either 250 or 1,000 ppm. After 14 days inhalation, they found slight cytoplasmic alterations at concentrations as low as 250 ppm. Hepatocyte necrosis with inflammatory reactions and Kupfer's cell hypertrophy were observed in the group exposed to 1,000 ppm [113]. Hepatotoxicity can be somewhat potentiated by enzyme induction with phenobarbital, 3-methyl-cholanthrene, and other inducers [114]. No distinct effects could be produced by induction with ethanol [115]. 1,1,1-Trichloroethane, however, has proved to be a potent enzyme inducer. The metabolism of short-acting barbiturates is increased in rats and mice after inhalation of 2,000 ppm and, therefore, the sleep period is shortened. In-vitro tests have shown the oxidation of barbiturates, and the demethylation of aminopyrene in liver extracts occur more rapidly after inhalation of 3,000 ppm of 1,1,1-trichloroethane [116, 117]. The induction effect of 1,1,1-trichloroethane can be inhibited by cyclohexamide and actinomycin, two cytostatically active protein synthesis inhibitors. Lower concentrations of 500 ppm did not produce enzyme induction after 7 days inhalation [118]. No intensifying effect was observed with nicotinic hydrochloride [119].

In animal experiments, 1,1,1-trichloroethane produced a decline in blood pressure in the left ventricle and major arteries, as well as cardiac arrhythmia [120]. Application of low doses (0.25 to 0.4 cm^3/kg body weight) and subsequent injection of adrenalin in anesthesized dogs resulted in ventricular extrasystoles and ventricular tachycardia [121, 122]. The thesis that cardiac arrhythmia is always induced by excretion of physiological amounts of catecholamine in connection with high inhalation doses of 1,1,1-trichloroethane was supported by experiments in which adrenal glands of mice had been removed prior to administration of different adrenalin doses [123]. Extensive studies with dogs and rabbits have dealt with pressure conditions in the heart and general circulation. At anesthetic concentrations, peripheral vasodilation as well as positively chronotropic and inotropic heart action develop initially; approximately 1 min later the stroke volume decreases and muscular tension is reduced. The indirect cardioactive effect in this second phase can be prevented by administration of calcium ions [124].

Kidney damage, predominantly swelling of the proximal tubules, develops at an anesthetic concentration equal to the LD_{80}. Necrosis or measurably impaired renal function was not observed [115, 125]. 1,1,1-Trichloroethane also appears to have an influence on the immune system. In rats and rabbits exposed for several month to concentrations as low as 18 ppm, the immune response to administration of typhoid vaccine was considerably weaker than that of the control group [115].

– *Human Toxicity*

1,1,1-Trichloroethane has a mild irritating effect on the skin and mucosa, starting at approximately 500 ppm [112]. Psychomotoric performance was significantly impaired at concentrations exceeding 350 ppm [126, 127]. Slight disturbances in coordination and pronounced dizziness developed at 900 ppm. Gait disturbances were pronounced at 1,900 ppm: test subjects were unable to stand unsupported at 2,600 ppm. Anesthetic concentrations range from 10,000 to 26,000 ppm [100, 112]. Deaths described in literature predominantly involved sniffers or were due to improper use in the home or at the workplace [112, 128, 129, 130, 131, 132, 133]. In most cases death was due to respiratory paralysis; in a few cases, death was caused by cardiac arrhythmia and ventricular fibrillation. Minimum human lethal concentration limits are not known.

Because of the low metabolic rate, specific organ damage is extremely unlikely. Acute intoxication is completely reversible. A few cases of reversible liver and kidney damage and one case of chronic liver damage have been described after acute intoxication-induced icteric necrosis [134, 135]. Occupational intoxication in chronically exposed workers has not been reported. One epidemiological study showed no significant differences between exposed and unexposed workers [136].

– Mutagenicity

The mutagenic effect of 1,1,1-trichloroethane was weak in S. typhimurium TA 100, with and without microsomal activation system [137]. 1,1,1-Trichloroethane induces in vitro transformation in the Fisher rat embryo cell system [138].

– Carcinogenicity

Tumor incidence in mice and rats fed with various doses of 1,1,1-trichloroethane or in rats after inhalation tests and intratracheal instillation did not differ significantly from that of the control group. Because of the high death rate of the experimental animals, however, definite conclusions are not possible at present. Further studies are in progress [139, 140].

– Teratogenicity

The teratogenic effect of 1,1,1-trichloroethane on chick embryos was more potent than that of other chlorinated hydrocarbons examined [141]. This finding, however, was not confirmed by another study in which pregnant rats and mice were exposed to 875 ppm during days 6 to 15 of gestation for 7 h a day. No damage of any kind was demonstrable in the brood or the embryos [34].

Sampling and Analysis

Air specimens were trapped with activated charcoal, desorbed, and analyzed with GC/FID, or GC/MS [78, 142]. The most reliable method for analyzing water samples is GC/MS [80].

Environmental Fate

– *General Environment*

At present, consumers are more responsible for the environmental load of 1,1,1-trichloroethane, than producers.

The substance is relatively unstable. In the atmosphere it decomposes rapidly by reacting with atmospheric hydroxyl radicals, ozone, and other substances. The tropospheric half-life is estimated to be 380 days [20]; the atmospheric lifetime is estimated to be 2 to 5 or 11 years [21, 143]. Measurements in North America indicated tropospheric concentrations of 145 ± 25 ppt; values in the lower stratosphere have been reported to be about 50% smaller. Concentrations of 65 ppt have been measured in the southern hemisphere [143]. At least 15% of the 1,1,1-trichloroethane in the atmosphere enters the stratosphere, were it comes into contact with the ozone layer. 1,1,1-trichloroethane is capable of decomposing 20% as much ozone as Freon 11 and 12, i.e., 1% of the stratospheric ozone is decomposed by 1,1,1-trichloroethane [20, 144, 145].

Mean air concentrations for 14 cities and rural areas in the United States ranged between 77 ppt and 0.83 ppt. Similar measurements were obtained in the municipal area of Tokyo, the Republic of Ireland, and South Africa. The 1,1,1-trichloroethane concentration varied markedly during the course of the day and depended on the amount of rain. It declined markedly on sunny days [142, 146, 147, 148, 149, 150].

A British study of six different towns showed that the air concentration (1 to 16 ppb) was significant higher than those of the studies cited above [109]. According to this study, rain water concentrations were as high as 90 mg/liter and tap water concentrations (including carbon tetrachloride) as high as 300 mg/liter. Seawater concentrations (3.3 μg/liter) were even higher; marine sediments contained as much as 5 μg/kg. In a cross-sectional study, carried out between 1972 and 1976 in several west European countries, mean concentrations ranged from 0.1 to 3.0 ppb. The distribution range for river, lake, and canal water measured at the same time was considerably broader [151]. Concentrations as 16.5 μg/liter were measured in untreated waste water. One group of researchers compared soil from farms and forests far from any industrial estate, land at some distance from industrial plants, and areas in the immediate vicinity of industrial plants. Mean concentrations of 1, 2.2, and 9.0 μg/m^3, respectively, were found at the three sites at drilling depths of 15 to 100 cm [152]. The same study established dose-dependent leaf damage, when the soil was gassed under standardized conditions with high doses of a mixture of 1,1,1-trichloroethane, trichloroethylene, and tetrachloroethylene. Doses were 1,000 to 3,000 times higher than previously detected atmospheric concentrations. 1,1,1-trichloroethane concentrations ranging from 1 to 10 mg/kg were reported in 12 food samples [144]. And it was consistently found in many aquatic plants and animals.

Table 8. Permissible exposure standards of 1,1,1-trichloroethane

Index	Permissible exposure ppm (mg/m^3)	Reference
TLV-TWA (ACGIH)	350 (1,900)	[5]
TLV-STEL (ACGIH)	450 (2,450)	[5]
TLV-C (ACGIH)	not adopted	[5]
BEI (ACGIH)	not adopted, but intended	[5]
	– 1,1,1-trichloroethane in exhaled air	
	– prior to the last shift of working week: 40 ppm	
	– Trichloroacetic acid in urine	
	– end of working week 10mg/L	
	– total trichloroethanol in urine	
	– end of shift at end of working week: 30 mg/L	
	– total trichloroethanol in blood:	
	– end of shift at end of working week: 1 mg/L	
PEL (OSHA)	350 (1,900)	[3]
(Final rule limit: TWA: 350 (1,900);		
STEL: 450 (2,450)		
REL (NIOSH)	350 (1,900)	[4]
MAK (GERMANY)	200 (1,080)	[24]
EPA criteria:		[25]
freshwater aquatic life:	5,300 µg/liter (24 h average)	
	12,000 µg/liter (MCC)	
saltwater aquatic life:	240 µg/liter (24 h average)	
	540 µg/liter (MCC)	
human health:	15.7 mg/liter	

– Workplace Environment

NIOSH estimates approximately 2.9 million workers in the United States are exposed to 1,1,1-trichloroethane [23]. Because of its broad application as a solvent, most of these workers are probably exposed to higher doses. High risk occupations are cleaning and degreasing jobs in the metal and textile industries as well as the production and use of aerosol cans containing 1,1,1-trichloro-ethane. Decomposition to phosgene in the presence of an open flame, hot metals, and lighted cigarettes is a special hazard. Experimental tests indicated that phosgene was also formed by shortwave radiation during inert gas shielded arc welding [153]. Stabilizing agents accumulate when 1,1,1-trichloroethane is used in vapor degreasing [112].

1,1,2-Trichloroethane

$$
\begin{array}{ccc}
\text{H} & & \text{H} \\
| & & | \\
\text{Cl–C–} & & \text{C–Cl} \\
| & & | \\
\text{Cl} & & \text{H}
\end{array}
$$

Synonyms

β-Trichloroethane, ethane trichloride, vinyl trichloride.

History

1,1,2-Trichloroethane does not occur naturally. The first synthesis was described by Regnault in 1840. Large scale production started in 1940 [2, 6].

Physical Properties

1,1,2-Trichloroethane is a colorless nonflammable liquid of pleasant sweet or chloroformlike odor which is miscible with most organic solvents and forms azeotropes with methanol, tetrachloroethylene, ethanol, and water. In the presence of air, 1,1,2-trichloroethane does not form ignitable mixtures [2, 6, 7, 22, 93].

Chemical Properties

With the exclusion of air and water, 1,1,2-trichloroethane is stable at temperatures below 100 °C. It is easily dehydrochlorinated to almost equal amounts of 1,1-dichloroethylene and 1,2-dichloroethylene and hydrogen chloride at temperatures from 400 °C to 500 °C. The process is accelerated by catalysts. Boiling water produces vigorous hydrolysis. Hazardous decomposition products are toxic gases and vapors such as hydrogen chloride, phosgene, and carbon monoxide. 1,1,2-Trichloroethane is highly corrosive to aluminium, iron, and zinc. Addition of water increases the corrosion rate [2, 6, 7, 22].

Production, Storage, and Transportation

The most important and economical production process is the ethylene induced selective chlorination of 1,2-dichloroethane. Various other processes are also employed, but they are not of practical importance.

Based on production figures for 1,1,1-trichloroethane and 1,1-dichloroethylene its western world production in 1984 is estimated to be 200,000 to 220,000 tons [2, 22].

Use

Most 1,1,2-trichloroethane is used as an intermediate in the production of 1,1-dichloroethylene and 1,1,1-trichloroethane. Its relative high toxicity does

Table 9. Physical properties of 1,1,2-trichloroethane

Molecular weight:	133.41
Density:	1,141 kg/m^3 (20 °C)
Vapor density:	4 kg/m^3 (1 bar, boiling point)
Vapor pressure:	48 mbar (30 °C)
	492 mbar (90 °C)
	906 mbar (110 °C)
Melting point:	-37 °C
Boiling point:	113 °C (1 bar)
Solubility in water:	0.45 % (20 °C)
Solubility of water in 1,1,2-trichloroethane:	0.05% (20 °C)
Log(octanol/water partition coefficient):	2.17
1 ppm = 5.548 mg/m^3	1 mg/m^3 = 0.108 ppm

not allow general use as a solvent, though it is used in smaller amounts in adhesives, the production of teflon tubing, in lacquer, and coating formulations [2, 22].

Pharmacokinetics

1,1,2-Trichloroethane may be inhaled, ingested or absorbed through the skin. Partition coefficients are: blood/air 38.6, and oil/blood, 59 [27]. After intraperitoneal administration of 0.1 to 0.2 g/kg body weight of labelled 1,1,2-trichloroethane in mice, 73 to 87% of activity were found in the urine and 16 to 22% in the respiratory air (40% unaltered, 60% carbon dioxide). Metabolites detected in urine were chloroacetic acid, S-carboxymethyl cysteine, and thiodiacetic acid. Small amounts of glycolic acid, dichloroethanol, trichloroethanol, oxalic acid, and trichloroacetic acid indicated 1,1,2-trichloroethane metabolism by formation of chloroacetaldehyde [154]. In the presence of liver microsomes, NADPH, and oxygen, enzymatic dechlorination could be demonstrated in vitro. This was accelerated by pretreatment with enzyme inducers, phenobarbitone, and benzopyrene [98]. Human pharmacokinetic studies have not been published.

Biomedical, Effects and Toxicity

– Animal Toxicity

Oral LD$_{50}$ in rats is 835 mg/kg body weight; intraperitoneal LD$_{50}$ in mice is 500 mg/kg body weight [30, 111]. Lethal inhalation concentrations after 3 and 7 h exposures were 18,000 and 14,000 ppm, respectively [110]. Percutaneous uptake was extremely high in guinea pigs [155]. Dose-dependent liver and kidney damage was demonstrated in different experimental animals with all methods of application. The hepatotoxicity of 1,1,2-trichloroethane was lower than that of chloroform and carbon tetrachloride, but higher than that of 1,1,1-trichloroethane. Pretreatment with enzyme inducers or simultaneous administration of polychlorinated biphenyls and experimental alloxan diabetes intensifies the hepatotoxic effect [111, 114, 155, 156, 157, 158].

– Human Toxicity

Lethal or anesthetic human concentrations are not known. There are no occupational or epidemiological studies on exposed groups available.

– Mutagenicity

1,1,2-Trichloroethane was not mutagenic in *S. typhimurium* TA 1535, with and without liver microsomes [68].

– Cancerogenicity

After feeding different doses of 1,1,2-trichloroethane to B6C3F1 mice, approximately 50% of the mice in each group were still alive 90 weeks later. The incidence of hepatocellular carcinomas in the treated animals was significantly higher than that in the control group. No significant differences however, were found in a similar experiment, using Osborne-Mendel rats [33].

– Teratogenicity

Experiments with fish and aquatic interverterbrates indicated that 1,1,2-trichloroethane impaired reproductivity [159],
    Other studies about teratogenicity or embryotoxicity are not available.

Sampling and Analysis

Air samples were trapped with activated charcoal and analyzed with GC/FID [78]. Most reliable water analyses are obtained with GC/MS [9].

Environmental Fate

– General Environment

Due to the low production rate of 1,1,2-trichloroethane, major air contamination is highly unlikely: measurements, however, are not available. When released to the atmosphere, 1,1,2-trichloroethane should degrade by reacting with hydroxylic radicals with a half-life of 24 days. The half-life in polluted atmospheres is much shorter, being of the order of 16 hours [22]. Concentrations of

Table 10. Permissible exposure standards of 1,1,2-trichloroethane

Index	Permissible exposure ppm (mg/m^3)	Reference
TLV-TWA (ACGIH)	10 (45)	[5]
TLV-STEL (ACGIH)	not adopted	
TLV-C (ACGIH)	not adopted	
BEI (ACGIH)	not adopted	
PEL (OSHA)	10 (45)	[3]
(Final rule limit: TWA: 10 (45); STEL: –; Ceil: –)		
REL (NIOSH)	10 (45)	[4]
MAK (GERMANY)	10 (45)	[24]
EPA criteria:		[25]
freshwater aquatic life:	310 μg/liter (24 h average)	
	710 μg/liter (MCC)	
saltwater aquatic life:	not established	
human health:	2.7 μg/liter	

5.4 mg/liter were obtained in industrial waste water. Concentrations in several random samples of drinking water in the United States and in Europe ranged from 0.1 to 8.5 ug/liter [9, 22].

– Workplace Environment

NIOSH estimates that 112,000 persons are exposed to 1,1,2-trichloroethane at their workplace in the U.S.A. There is no information available on mean exposure concentrations. The highest concentrations were found in individuals working with blast furnaces, in steel rolling mills, and in factories manufacturing technical instruments [23].

1,1,1,2-Tetrachloroethane

$$
\begin{array}{ccc}
\text{Cl} & & \text{H} \\
| & & | \\
\text{Cl–C–} & \text{C–Cl} \\
| & & | \\
\text{Cl} & & \text{H}
\end{array}
$$

Synonym

asym-Tetrachloroethane

History

1,1,1,2-Tetrachloroethane does not occur naturally. It was first synthesized by A. Mouneyrat in 1898. Today it may be found as a byproduct in several industrial chlorination reactions of C_2 hydrocarbons. It is not produced on an industrial scale. Because of its high toxicity, it is not used as a solvent [2].

Physical Properties

1,1,1,2-Tetrachloroethane is a white to yellowish red, nonflammable, heavy liquid [2, 6].

Chemical Properties

1,1,1,2-Tetrachloroethane is chemically more stable than 1,1,2,2-tetrachloroethane its isomer. Thermal decomposition to trichloroethylene and hydrogen chloride starts at 500 °C. Tetrachloroethylene can be formed by disproportionation. Dichloroacetyl chloride is obtained through oxidation [2, 6].

Production, Storage, and Transportation

Pure 1,1,1,2-tetrachloroethane is produced by the addition of chlorine to 1,1-dichloroethylene. It is a byproduct of trichloroethylene produced from ethylene. The substance is not commercially produced in the U.S.A [2, 6, 22].

Use

1,1,1,2-Tetrachloroethane is an undesired byproduct mainly in the production of 1,1,1-trichloroethane from 1,1-dichloroethane, 1,1,2-trichloroethane and 1,1,2,2-tetrachloroethane from 1,2-dichloroethane. Its most important commercial use is its conversion to tetrachloroethylene in the chlorinolysis process. Small amounts are used in manufacturing insecticides, herbicides, soil fumigants, bleaches, paints, and varnishes [2, 6, 22].

Pharmacokinetics

Partition coefficients are: blood/air 30.4, and oil/blood, 142 [27]. Subcutaneous administration to mice showed half of the dose exhaled unchanged, metabolized

Table 11. Physical properties of 1,1,1,2-tetrachloroethane

Molecular weight:	167.68
Density:	1.54 (20 °C)
Vapor pressure:	10 mmHg (19.3 °C)
Melting point:	− 68 °C
Boiling point:	129 °C (1.013 bar)
Heat of Vaporization:	9296.5 gcal/gmole

parts were excreted mainly as trichloroethanol and to a minor extent as trichloroacetic acid. Inhalation or ingestion of 1,1,1,2-tetrachloroethane by pregnant rats and rabbits resulted in presence of high levels in fetuses, indicating trans-placental passage [22].

Biomedical Effects and Toxicity

– Animal Toxicity

Tests with mice, rats, rabbits and hares indicate only a slight irritant effect of 1,1,1,2-tetrachloroethane on the skin and mucosa. Percutaneous absorption is low. 1,1,1,2-tetrachloroethane is two to three times less toxic than 1,1,2,2-tetrachloroethane. Histological findings were microvacuolization and centrilobular necrosis in the liver. Slight alterations were also found in the myocardium. 1,1,1,2-tetrachloroethane (100–800 mMol/kg/Day administerred intraperitoneally for seven days in male rats, increased liver succinate dehydrogenase, acid phosphatase, and glucose-6-phosphatase activities. Liver DNA content decreased and white blood cell count increased, while red blood cell count and blood cholesterol decreased [160].

– Human Toxicity

There are no reports about acute or chronic intoxication in man. Epidemiological studies are not available.

– Mutagenicity

1,1,1,2-Tetrachloroethane has no mutagenic effect in S. typhimurium strains TA 98, TA 100, TA 1535, TA 1537, and TA 1538 with and without Aroclor-induced S9 [22].

– Carcinogenicity

An increase of hepatocellular carcinomas in females, and an increase of hepatocellular adenomas in female and male B6C3F1 mice have been observed, when fed with 1,1,1,2-tetrachloroethane. Fisher 344/N rats showed no carcinogenic effects, although the observed increase of liver tumors in male rats may be associated with the administration of the substance.

– Teratogenicity

Although 1,1,1,2-tetrachloroethane passes the placental barrier, no productive disorders were observed.

The offspring of brood animals treated with 1,1,1,2-tetrachloroethane died within 2 days after birth [161, 162, 163].

Sampling and Analysis

Gas chromatography analysis, using the purge-and-trap method, was used for the detection of 1,1,1,2-tetrachloroethane. Other analytical techniques such as TLC and GC/MS have been reviewed [22].

Environmental Fate

– General Environment

Since its production is low, 1,1,1,2-tetrachloroethane is not an environmental hazard. Measurement data are not available, but field studies were conducted in Phoenix, AZ, Los Angeles, and Oakland, CA, to better characterize the atmospheric abundance, fate and human exposure of selected organic chemicals that may be potentially hazardous. Concentrations, variabilities and average daily dosages from exposure to the haloethanes, including 1,1,1,2-tetrachloroethane had been determined to be 142 mg/day [150].

– Workplace Environment

In the U.S.A and Europe, there should be no workplaces where 1,1,1,2-tetrachloroethane is in regular use.

Table 12. Permissible exposure standards of 1,1,1,2-tetrachloroethane

Index	Permissible exposure ppm (mg/m^3)	Reference
TLV-TWA (ACGIH)	not adopted (under study)	[5]
TLV-STEL (ACGIH)	not adopted	
TLV-C (ACGIH)	not adopted	
BEI (ACGIH)	not adopted	
PEL (OSHA)	not established	[3]
REL (NIOSH)	not recommended	[4]
MAK (GERMANY)	not established	[24]
EPA criteria:		[25]
freshwater aquatic life:	420 µg/liter (24 h average) 960 µg/liter (MCC)	
saltwater aquatic life:	not established	
human health:	not established	

1,1,2,2-Tetrachloroethane

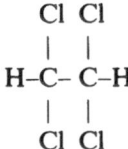

Synonyms

Acetylene tetrachloride, sym-tetrachloroethane, 1,1-dichloro-2,2-dichloro-ethane, ethane tetrachloride.

History

1,1,2,2-Tetrachloroethane does not occur naturally. The first synthesis by M. Berthelot and E. Jungfleisch was reported in 1869. Based on experiments by A. Mouneyrat, commercial production was started by A. Wacker in 1903.

1,1,2,2-Tetrachloroethane became the first chloroethane to be produced in large quantities.

It was once used as a solvent, but increasing rates of liver injuries with a lot of fatal cases led to a rapid decline in its use. In most countries limitation in use are recommended by the government because of its high toxicity since early 1920th [2, 6].

Physical Properties

1,1,2,2-Tetrachloroethane is a colorless, dense, nonflammable liquid of penetrating, sweet, chloroformlike odor. It is miscible with all common organic solvents and has the major solvent power of all aliphatic chlorinated hydrocarbons. It forms no explosive mixtures with air [2, 6]. The odor threshold is less than 3 ppm [5].

Chemical Properties

In the absence of light, air, and moisture, 1,1,2,2-tetrachloroethane is stable. It decomposes to trichloroethylene and hydrogen chloride at temperatures above 400 °C. This process is demonstrable with suitable catalytic crackers at temperatures as low as 250 °C. In the atmosphere it decomposes to hydrogen chloride and small amounts of phosgene. Dichloroacetyl chloride and phosgene are formed by photooxidation. 1,1,2,2-tetrachloroethane is resistant to strong acids. Weak alkalies split off trichloroethylene; strong alkalies split off explosive dichloroacetylene, one of the most neurotoxic agents. Iron, and zinc reduce 1,1,2,2-tetrachloroethane to 1,2-dichloroethane in the presence of steam. The substance can be stored without adding stabilizers [2, 6, 22].

Production, Transportation, and Storage

There are two main routes of industrial production of 1,1,2,2-tetrachloroethane:

1. Addition of chlorine to acetylene, and
2. the liquid phase chlorination of ethylene or the ethylene induced chlorination of 1,2-dichloroethane, which is the same process as that used in the production of 1,1,2-trichloroethane. The technical yield of the first process is 90 to 98%, but due to the high costs of acetylene, induced chlorination of ethylene or 1,2-dichloroethane has recently become more important.

Several other processes have been patented, but all of them are only of minor importance. The degree of purity of 1,1,2,2-tetrachloroethane is 95–98%. It is stored and transported in iron tanks, it can be stored without adding stabilizers [2, 6, 22].

Use

1,1,2,2-Tetrachloroethane is almost always used as an intermediate in the production of trichloroethylene. As a solvent it is very rarely used because of its very high toxicity. Other uses are as a mothproofing agent, an insecticide in greenhouses, and as grain disinfectant [2].

Pharmacokinetics

1,1,2,2-Tetrachloroethane is primarily absorbed via the lungs; percutaneous absorption is possible, but slight. The partition coefficients are: blood/air, 121.4,

Table 13. Physical properties of 1,1,2,2-tetrachloroethane

Molecular weight:	167.86
Density:	1,597 kg/m^3
Vapor density:	5 kg/m^3 (boiling point, 1 bar)
Vapor pressure:	7 mbar (20 °C)
	53 mbar (60 °C)
	187 mbar (91 °C)
	827 mbar (138 °C)
Melting point:	− 42.5 °C
Boiling point:	146.35 °C
Solubility in water:	0.29 mass% (20 °C)
Solubility of water in 1,1,2,2-tetrachloroethane:	
	0.03 mass% (20 °C)
Log(octanol/water partition coefficient):	2.56
1 ppm = 6.976 mg/m^3	1 mg/m^3 = 0.134 ppm

and oil/blood, 109 [27]. Labelled 1,1,2,2-tetrachloroethane was eliminated within 3 days after intraperitoneal injection: 45 – 61% exhaled as carbon dioxide; less
than 4% was exhaled unaltered; 23 – 34% was excreted in the urine, and 16% of
the injected dose remained in the body. Dichloroacetic acid, trichloroacetic acid,
trichloroethanol, and oxalic acid were detected as metabolites in the urine [164].
After 8 h inhalation of 200 ppm, trichloroacetic acid and trichloroethanol were
also excreted in the urine. Comparative volunteer tests with other labelled
halogenated hydrocarbons showed, with short-term inhalation, the lowest elimination rate of all compounds was that of 1,1,2,2-tetrachloroethane: only approximately 3% had been exhaled within one hour [101].

Biomedical Effects and Toxicity

– Animal Toxicity

The LC_{50} in male rats after 4 h inhalation was 8.6 mg/liter [164]. The oral LD_{50}
in rats was 250 mg/kg body weight; the intraperitoneal LD_{50} in mice was
820 mg/kg body weight [162, 165]. 1,1,2,2-Tetrachloroethane is a strong irritant
to the skin and mucosa. Maximum anesthetic concentrations in experimental
animals are not known. 1,1,2,2-Tetrachloroethane is a very potent hepatotoxic
substance. After just 3 h inhalation of 800 ppm, triglycerides and phospholipides
increased significantly in mouse livers; maximum concentration was attained
after 25 h. The alterations, however, were not as severe as those developing after
carbon tetrachloride exposure. Benzopyrene hydroxylase and p-nitroanisole-
O-demethylase concentrations declined 50% within 24 h after feeding one single
dose of 437 mg/kg body weight to rats [166]. After long-term exposure, inflammatory alterations were detectable in liver. This effect was intensified by raising
the ambient temperature. With high doses, mortality rate in mice depends on
time of day, too. Fatty degeneration of renal tubules as well as icteric necrosis
was found after acute intoxication [167, 168].

– Human Toxicity

1,1,2,2-Tetrachloroethane is a strong irritant of the skin and mucosa and
a potent anesthetic. Minimum anesthetic doses in man have not been established.

1,1,2,2-Tetrachloroethane causes nephritis and severe toxic hepatitis with
acute yellow atrophy of the liver. If the exposure is severe, within a few hours
a deep dusky coloration of the skin may appear. 1,1,2,2-Tetrachloroethane may
also cause some central nervous system depression. In very severe acute exposures unconsciousness and death from respiratory failure have been seen. Eight
people, each of whom ingested 3 ml by mistake, became comatose and arreflexic,
but recovered without sequelae. Chronic intoxication by 1,1,2,2-tetrachloro-
ethane can take two forms: Central nervous system effects, such as tremor,

vertigo, and headache; and gastrointestinal and hepatic symptoms with nausea, vomiting, gastric pain, jaundice, and enlargement of the liver [22]. Many cases of severe or fatal intoxication resulting from accidental inhalation or ingestion at the workplace were reported prior to 1930. Pronounced to extremely severe toxic alterations of the liver and toxic fatty degeneration of the renal tubules were always present in such cases. Severe cerebral damage and peripheral neuroparalysis have also been reported. Alterations were observed after long-term exposure. All of these reports lack information about exposure levels [169, 170, 171]. Recent epidemiological studies are not available.

– Mutagenicity

1,1,2,2-Tetrachloroethane had a mutagenic effect on *S. typhimurium* strains TA 1530 and TA 1535, but not on TA 1538. Mutagenic and cytotoxic effects were also demonstrated in other bacterial systems [64, 172].

– Carcinogenicity

The results of ingestion tests showed a significant dose-dependent incidence of hepatocellular carcinomas in mice. The same tests in rats, however, showed no significant difference between treated and control groups [33]. The difference between mice recieving an intraperitoneal injection and the control group was not significant. Since the death rate of the animals was high, these findings should be regarded with reservation [70].

– Teratogenicity

Deformities were observed in embryos and young animals after treatment of pregnant mice daily with 1,1,2,2-tetrachloroethane doses of 300 mg/kg body weight [173].

Sampling and Analysis

Air samples were trapped with activated charcoal, desorbed, and analyzed with GC/FID [78]. Most reliable water analyses are provided by GC/MS [79].

Environmental Fate

– General Environment

Most of the 1,1,2,2-tetrachloroethane released enters the environment during production and utilization. In Japanese cities, 1,1,2,2-tetrachloroethane concentrations of 0.01 – 9.4 ppb have been detected [87]. In the United States and

Table 14. Permissible exposure standards of 1,1,2,2-tetrachloroethane

Index	Permissible exposure	Reference
	ppm (mg/m³)	
TLV-TWA (ACGIH)	1 (7)	[5]
TLV-STEL (ACGIH)	not adopted	
TLV-C (ACGIH)	not adopted	
BEI (ACGIH)	not adopted	
PEL (OSHA)	5 (35)	[3]
(Final rule limit: TWA: 1 (7); STEL: – ; Ceil: –)		
REL (NIOSH)	5 (35)	[4]
MAK (GERMANY)	1 (7)	[24]
EPA criteria:		[25]
freshwater aquatic life:	170 μg/liter (24 h average)	
	380 μg/liter (MCC)	
saltwater aquatic life:	70 μg/liter (24 h average)	
	160 μg/liter (MCC)	
human health:	1.8 μg/ml	

Europe, concentrations of 0.01–0.11 μg/liter were obtained in drinking water, and 2.2 mg/liter in industrial waste water [20, 174]. 1,1,2,2-Tetrachloroethane has also been found in industrial vinyl chloride wastes, dumped in the North Sea [88].

In the air it is extremly stable, with a half-life of more than two years. Some of the chemical will eventually diffuse into the stratosphere, where it will rapidly photodegrade. 1,1,2,2-Tetrachloroethane, which is released into water, will primarily be lost by volatilization in a matter of days to weeks. When disposed on soil, parts of it may leach into the ground water. There is evidence of slow biodegradation. Under anaerobic conditions one product of biodegradation is 1,1,2-trichloroethane, which is resistent to furthr biodegradation. 1,1,2,2-Tetrachloroethane is not expected to concentrate in the food chain. Major source of human exposure is from ambient air or drinking water near industrial sites.

– Workplace Environment

NIOSH estimates that approximately 11,000 persons are occupationally exposed to 1,1,2,2-tetrachloroethane [23]. Production, storage, transport, use, and disposal represent a hazard to those people involved. There are no statistics available on the number of exposed individuals.

Pentachloroethane

$$
\begin{array}{ccc}
 & Cl & H \\
 & | & | \\
Cl-C- & C- & Cl \\
 & | & | \\
 & Cl & Cl
\end{array}
$$

Synonym

Ethane pentachloride.

History

Pentachloroethane does not occur naturally. It was first synthesized by V. Regnault in 1840 by chlorination of monochloroethane. In the first half of 20th century it was produced as an intermediate for the tetrachloroethylene process (Pentachloroethane pyrolysis) [2, 6].

Physical Properties

Pentachloroethane is a colorless, dense, nonflammable liquid with a chloroformlike sweetish odor. It does not form explosive mixtures with air and it is soluble in all common solvents [2, 6].

Chemical Properties

In the absence of air and moisture,pentachloroethane is stable. In the presence of water, it hydrolyzes at ambient temperatures. It is decomposed by weak alkalies or heat to tetrachloroethylene and hydrogen chloride with small amounts of trichloroethylene and chlorine. Dehalogenation by reaction with alkalies or metals will produce spontaneously explosive chloroacetylenes. Dichloroacetyl chloride is formed with fuming sulfuric acid. Air oxidation in the presence of UV light gives trichloroacetyl chloride [2, 6, 7].

Production, Storage, and Transportation

Pentachloroethane can be produced by the addition of chlorine to trichloroethylene using ferric chloride as a catalyst and by induced chlorination of 1,2-dichloroethane with ethylene. Pentachloroethane must be stabilized. After being stabilized, it can be stored in iron tanks for extended periods of time. Since tetrachloroethylene production is more economic by the chlorinolysis process, industrial production of pentachloroethane has become unimportant and is no longer used [2, 6, 7].

Use

Pentachloroethane was exclusively used as an intermediate product of tetrachloroethylene production. Because of its low stability and its high toxicity, the

Table 15. Physical properties of pentachloroethane

Molecular weight:	202.31
Density:	1,680 kg/m³ (20 °C)
Vapor density:	6 kg/m³
Vapor pressure:	4.5 mbar (20 °C)
	35 mbar (60 °C)
	79 mbar (80 °C)
	173 mbar (100 °C)
Melting point:	-29 °C
Boiling point:	161 °C
Solubility in water:	0.05 mass% (20 °C)
Solubility of water in pentachloroethane:	0.24 mass% (20 °C)
1 ppm = 8.4088 mg/m³	1 mg/m³ = 0.119 ppm

use of pentachloroethane as a solvent for cellulose derivates, rubbers, and resins is not significant, but may occur [2].

Pharmacokinetics

There are no studies of uptake, distribution, metabolism and excretion of pentachloroethane in man.

About one third of the pentachloroethane administered to mice was exhaled unchanged via the lungs. About one third was excreted as 2,2,2-trichloroethanol and about 9 to 18% as trichloroacetic acid in the urine. Indicating both dechlorination and hydrochlorination of pentachloroethane, another study found trichloroethylene and tetrachloroethylene in expired air of mice [22].

Biomedical Effects and Toxicity

- Animal Toxicity

Pentachloroethane, one of the most toxic chloroethanes, is a strong irritant to skin and mucosa. Its a more potent anesthetic than chloroform. The LD_{50} in dogs after intravenous injection is 100 mg/kg body weight and the inhalatory LD_{50} is 35 mg/liter [28, 175]. Anesthetic doses range from 13 to 37 mg/liter, depending on the duration of anesthesia. Inhalation of 120 ppm for 8 h a day over a period of 23 days was tolerated by cats without any signs of CNS impairment [176]. Pronounced toxic alterations, however, were found in the liver, kidneys and lungs. Fatty degeneration of the liver and toxic alterations of kidneys and lungs were observed in dogs after 3 weeks exposure [177].

- Human Toxicity

Toxic damage after acute or chronic exposure has not been reported in man. Epidemiological studies are not available.

– Mutagenicity, Carcinogenicity, Teratogenicity

Studies on the mutagenicity, carcinogenicity, and teratogenicity in man have not been published. In four strains of *S. typhimurium* pentachloroethane was been mutagenic. Another study described a mutagenic potency, but there are no informations about the test itself. Pentachloroethane does not increase the SCE-rate. The substance showed significant increase of hepatocellular carcinoma in B6C3F1 mice and of tumors of the kidney in Fisher 344/N rats. How far this was influenced by the presence of an impurity in the form of about 4.2% Hexachloroethane is open to question. Hexachloroethane is reported to produce tumors in the liver and kidney [22].

Sampling and Analysis

Pentachloroethane in air can be analyzed by adsorption on poropak, desorption with hexane and then using GC/ECD [22].

Environmental Fate

– General Environment

Information about environmental load and fate is not available.

– Workplace Environment

Production of pentachloroethane and the occurrence of the substance as a by-product in chemical processes may be hazardous [2].

Table 16. Permissible exposure standards of pentachloroethane

Index	Permissible exposure ppm (mg/m^3)	Reference
TLV-TWA (ACGIH)	not adopted	[5]
TLV-STEL (ACGIH)	not adopted	
TLV-C (ACGIH)	not adopted	
BEI (ACGIH)	not adopted	
PEL (OSHA)	not established	[3]
REL (NIOSH)	not recommended	[4]
MAK (GERMANY)	5 (40)	[24]
EPA criteria:		[25]
freshwater aquatic life:	440 µg/liter (24 h average)	
	1,000 µg/liter (MCC)	
saltwater aquatic life:	38 µg/liter (24 h average)	
	87 µg/liter (MCC)	
human health:	not established	

Hexachloroethane

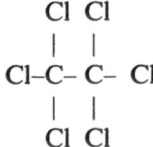

Synonyms

1,1,1,2,2,2-Hexachloroethane, hexachloroethylene, carbon hexachloride, ethane hexachloride, perchloroethane.

History

Hexachloroethane does not occur naturally. Commercial production started in the early 1920s [2, 6].

Physical Properties

At ambient temperatures, hexachloroethane is the only solid compound of all chlorinated C_2 alkanes and alkenes. Hexachloroethane forms colorless crystals of camphor-like odor. It occurs in three forms: rhombic (below 46 °C), triclinic (46–71 °C), and cubic (above 71 °C). It forms azeotropes with numerous organic compounds. Its physical properties are summarized below in Table 17 [2, 6].

Chemical Properties

When slowly warmed, hexachloroethane sublimes without melting or decomposition. Major amounts of chlorine are split off at temperatures above 250 °C. At higher temperatures it is resistant to acids and alkalies. In the presence of water, however, it corrodes iron. In the presence of metals, it dechlorinates to tetrachloroethylene and metal chlorides. Temperatures above 400 °C lead to disproportion to tetrachloroethylene and carbon tetrachloride. Spectral hexachloroethane decomposes to produce phosgene between 300 and 500 °C. Dehalogenation by reactions with alkalies, metals, etc., will produce spontaneous explosive chloroacetylenes [2, 6].

Production, Storage, and Transportation

Hexachloroethane is produced by chlorination of tetrachloroethylene with ferric chloride at temperatures of 100 to 140 °C [6]. There is no information available on impurities and stabilizing agents, or special requirements for storage and transportation. Production figures cannot be estimated.

Use

Because of its specific properties, such as the tendency to sublime and its very high chlorine content, hexachloroethane has some specific applications, which are limited, however, for toxicological and ecological reasons. It is primarily used for the production of smoke candles and grenades by mixing it with zinc, magnesium, and aluminium dust plus an oxygen donor. After ignition, this mixture forms fuming metallic chlorides. Small quantities are used for degassing in aluminium and magnesium smelting, as ignition supressant in flammable liquids, as a compound of high pressure lubricants, as a plasticizer in cellulose esters, as an additive in fire extinguishing agents, as a mothproofing agent, and as an antihelmintic in veterinary medicine [2, 6, 7].

Pharmacokinetics

Hexachloroethane may be absorbed from the gastrointestinal tract, through the lungs, and through the skin. Orally administered hexachloroethane is absorbed and appears rapidly in the systemic circulation. It is distributed widely throughout the body. The highest concentrations are found in fat, the lowest in muscle tissue [178]. Partition coefficients have not been established. Labelled hexachloroethane is slowly metabolized in rats: 14–24% of the radioactivity was exhaled as carbon dioxide, hexachloroethane, tetrachloroethylene, and 1,1,2,2-tetrachloroethane. Only 5% was excreted in the urine 3 days after ingestion.

The following metabolites of hexachloroethane have been identified: trichloroethanol (1.3%), dichloroethanol (0.4%), trichloroacetic acid (0.8%), monochloroacetic acid (0.7%), and oxalic acid (0.1%). The rest remained in the body [179]. Dechlorination of hexachloroethane was observed in rabbit liver

Table 17. Physical properties of hexachloroethane

Molecular weight:	236.76
Density:	2,091 kg/m^3 (20 °C)
Vapor density:	6.3 kg/m^3 (subl. pt., 1 bar)
Vapor pressure:	0.29 mbar (20 °C)
	1.33 mbar (40 °C)
	24 mbar (80 °C)
	116 mbar (120 °C)
Melting point:	188 °C
Boiling point:	185 °C
Solubility in water:	50 mass-ppm (20 °C)
Log(octanol/water partition coefficient):	3.34
1 ppm = 9.840 mg/m^3	1 mg/m^3 = 0.102 ppm

homogenates [180]. Formation of pentachloroethane was also detected in sheep [178]. There is no information about the fate of hexachloroethane in human body.

Biomedical Effects and Toxicity

– Animal Toxicity

Hexachloroethane is a mild irritant to the skin and mucosa. It is a more potent anesthetic than chloroform. The substance produces chronic liver lesions and acute diffuse nephrosis or digestive disorders in cattle. Severe inhalative intoxication in rats was observed at 5,000 ppm.

The LD_{50} in rodents is between 1,000 and 7,000 mg/kg [181]. Microsomal monooxygenase activity in rat liver declined 50% after administration of oral doses as low as 2.5 g/kg body weight. A substantial induction of hepatic enzymes, however, cannot be demonstrated [181]. Liver damage was observed at higher doses [178, 181].

– Human Toxicity

Worker exposure to fumes from hot hexachloroethane has been observed to cause blepharospasm, photophobia, lacrimation, reddening of the conjunctivae, but no corneal injury and no permanent damage. Concentration of exposure is not known. The substance shows more potent central nervous effects than chloroform or carbon tetrachloride, but is slower in action [22]. Other acute and chronic effects in man have not jet been described.

– Mutagenicity

Mutageñiciy tests in five strains of S. typhimurium and on Saccharomyces cerevisiae D4 showed no evidence of mutagenicity [22].

– Carcinogenicity

The incidence of dose-dependent hepatocellular carcinoma in B6C3F1 mice, fed various doses of hexachloroethane for 78 weeks was significantly higher than in the control group. This could not be demonstrated in Osborne-Mendel rats under similar test conditions. Due to the high death rate of experimental animals, however, no conclusions can be drawn from these tests [33].

– Teratogenicity

Teratogenic damage was not demonstrated in a study using Sprague Dawley rats [33].

Sampling and Analysis

Air samples were trapped with activated charcoal, desorbed, and analyzed with GC/FID [158]. The most reliable water analyses were obtained with GC/MS [90]. Solid waste analyses were conducted using chromatography and spectrochemical analysis.

Environmental Fate

– General Environment

No reports have been published on hexachloroethane detection in the atmosphere. It was, however, identified in waste water of paper mills (concentration < 1 μg/liter) and in samples of tap water in 4 from 13 cities (concentration of 0.03–4.3 μg/liter). Concentrations of river water and industrial waste water samples range from 4.4 to 8.4 ug/liter. Hexachloroethane could only be detected in surface water samples of highly industrialized areas [90, 182, 183].

– Workplace Environment

NIOSH estimates that approximately 1,500 workers have contact with hexachloroethane.
Paperboard mills are particularely hazardous [23].

Table 18. Permissible exposure standards of hexachloroethane

Index	Permissible exposure ppm (mg/m³)	Reference
TLV-TWA (ACGIH) intended change 1989: 1 (10)	10 (100)	[5]
TLV-STEL (ACGIH)	not adopted	
TLV-C (ACGIH)	not adopted	
BEI (ACGIH)	not adopted	
PEL (OSHA) (Final rule limit: TWA: 1 (10); STEL: –; Ceil: –)	1 (10)	[3]
REL (NIOSH)	not recommended	[4]
MAK (GERMANY)	1 (10)	[24]
EPA criteria:		[25]
freshwater aquatic life: 62 μg/liter (24 h average) 140 μg/liter (MCC)		
saltwater aquatic life: 7 μg/liter (24 h average) 16 μg/liter (MCC)		
human health: 5.9 μg/liter		

References

1. Flick EW (1985) Industrial solvents handbook, 3rd edn, Noyes, Park Ridge NJ
2. Rossberg M et al. (1986) Chlorinated hydrocarbons. In: Gerhartz W (ed) Ullmann's encyclopedia of industrial chemistry, 5th edn, vol A 6, VCH, Weinheim, Deerfield Beech, FL
3. US Department of Labor Occupational Safety and Health Administration (1989) OSHA final rule air contaminants permissible exposure limits. Title 29 code of Federal Regulations Part 1910.1000. Washington
4. US Department of Health and Human Services (1987) NIOSH pocket guide to chemical hazards. DHHS(NIOSH) Publ.No. 85–114. US Government Printing Office, Washington DC
5. American Conference of Governmental Industrial Hygienists (ACGIH) (1988) Threshold limit values and biological exposure indices for 1988–1989. Cincinnati, OH
6. Rassaert H, Witzel D (1975) Aliphatische Chlorkohlenwasserstoffe. In: Bartholome E et al. (eds) Ullmanns Encyklopädie der technischen Chemie, 4th edn, vol 9 Verlag Chemie, Weinheim, 404
7. Winnacker K, Küchler L (1972) Chemische Technologie. 3rd rev. edn., Vol 4: Organische Technologie II, München
8. Killian H, Weese H (1954) Die Narkose. Thieme, Stuttgart
9. Konietzko H (1984) Chlorinated Ethanes: Sources, distribution, environmental impact, and health effects. In: Saxena J (ed) Hazard assessment of chemicals: Current developments. vol 3, Academic, Orlando
10. Elfskind L (1928) Bruns Beitr Klin Chir 167: 25
11. Williams RT (1957) Detoxication Mechanisms. Chapman and Hall, London
12. Nuckolls AH (1933) Underwriter's Laboratory Rep No 2375 Cited in "The Halogenated Hydrocarbons. Toxicity and Potential Dangers" (ed.: Van Oettingen WF) (1955) Publ. Health Serv. Publ. NO. 414. Washington DC, U.S. Public Health Services, Washington DC
13. Sayers RR, Yant WP, Thomas BH, Bürger LB (1929) U.S. Publ Health Bull 185
14. Sayers RR, Yant WP, White CP, Patty FA (1930) Publ Health Rep 45: 225
15. Frey E (1912) Biochem J 40: 29
16. König R (1933) Langenbecks Arch Chir 99: 147
17. Davidson BM (1926) J Pharmacol Exp Ther 26: 37
18. Adrian J (1967) The Pharmacology of Anesthetic Drugs. Thomas, Springfield IL
19. Hes JP, Cohn DF, Streifler M (1979) Isr Ann Psychiatry Relat Discip 17: 122
20. Gäb S (1981) WABoLu-Berichte 3: 55
21. Chang JS, Penner JE (1978) Atmos Environ 12: 1867
22. Hazardous Substances Data Bank (HSDB). Established by National Library of Medicine (NLM). Part of TOXicology Data NETwork (TOXNET); Since 9th of August 1988 continuing "Toxicology Data Bank" (TDB)
23. National Institute for Occupational Safety and Health (NIOSH) (1978) Chloroethanes: Review of Toxicity. NIOSH Curr. Intell. Bull. No. 27. U.S. DHEW, Washington
24. Commission of the "Deutsche Forschungsgemeinschaft" for the Investigation of Health Hazards of Chemical Compounds in the Work Area (Ed.): Maximum Concentrations at the Workplace and Biological Tolerance Values for Working Materials 1989. Report No. XXV. Weinheim; VCI 1989. Newest German Edition: "Maximale Arbeitsplatzkonzentration und Biologische Arbeitsstofftoleranzwerte 1990." Mitteilung XXVI. (ed.: Senatskommission zur Prüfung gesundheitsschädlicher Arbeitsstoffe (Hrsg.). Deutsche Forschungsgemeinschaft. VCI, Weinheim 1990
25. US EPA (1982) Management of Hazardous Waste Leachate. EPA Contract No. 68-03—2766
26. Hansch C, Vittoria A, Silipo C, Jow PYC (1975) J Med Chem 18: 546
27. Sato A, Nakajima T (1979) Arch Environ Health 34: 69
28. Lazarew NW (1929) Arch Exp Pathol 141: 19
29. Müller J (1925) Arch Exp Pathol Pharmakol 109: 276
30. Smyth HF, Carpenter CP, Werl CS, et al. (1969) Am Ind Hyg Assoc J 30: 470
31. Spector WS (ed) (1956) Handbook of Toxicology. Saunders, Philadelphia
32. Hofmann HT, Birnstiel H, Jobst P (1970) Arch Exp Pathol Pharmakol 266: 360
33. National Cancer Institute (NCI) (1978) Bioassay of Hexachloroethane for Possible Carcinogenicity. DHEW Publ. No. (NIH) 78-1318. US Dept Health, Education, and Welfare, Washington DC
34. Schwetz BA, Leong BK, Gehring PJ (1974) Toxicol Appl Pharmacol 28: 452
35. Duerwald W (1954) Arch Toxicol 15: 144

36. Radding SB, Liu DH, Johnson HL, Mill T (1977) Review of the Environmental Fate of selected Chemicals. EPA-560/5-77-003. U.S. Environmental Protection Agency (Office of Toxic Substances), Washington DC
37. Gold LS (1980) Banbury Rep 5: 209
38. Fishbein L (1979) Potential industrial carcinogens and mutagens. Studies in environmental sciences. 4th Ed. Elsevier, Amsterdam, Oxford, New York
39. Reitz RH, Fox TR, Domoradzki JY, et al. (1980) Banbury Rep 5: 135
40. Yllner S (1971) Acta Pharmacol Toxicol 29: 471
41. Livesey JC, Anders MW (1979) Drug Metab Dispos 7: 199
42. International Agency for Research on Cancer (IARC) (1979) Some halogenated hydrocarbons. IARC Monographs on the evaluation of carcinogenic risk of chemicals to humans.vol. 20, Lyon
43. Gurney R (1943) Gastroenterology 1: 1112
44. Clayton GD, Clayton FE (Eds.) (1981–1982) Patty's Industrial Hygiene and Toxicology. vol 2A–C. 3rd Ed John Wiley Sons, New York
45. Spencer HC, Rowe VK, Adams EM, et al (1951) Arch Ind Hyg Occup Med 4: 482
46. Heppel LA, Neal PA, Petrin TL, et al (1946) J Ind Hyg Toxicol 28: 113
47. Kistler GH, Lückhardt AB (1929) Curr Res Anesth Analg 8: 65
48. Heppel LA, Neal PA, Endicott KM, Porterfield V (1944) Arch Ophthalmol 32: 391
49. Steindorff K (1922) Arch Ophthalmol 109: 253
50. Hinkel GK (1965) Dtsch Gesundheitswesen 20: 1327
51. Hueper WC, Smith C (1935) Am J Med Sci 189: 778
52. Reinfried H (1958) Dtsch Gesundheitswesen 13: 778
53. Weiss F (1957) Arch Gewerbepathol Gewerbehyg 15: 253
54. Wirtschafter ZI, Schwartz ED (1939) J Ind Hyg 21: 126
55. Brass K (1949) Dtsch Med Wochenschr 74: 553
56. Menschick H (1957) Arch Gewerbepathol Gewerbehyg 15: 241
57. Freundt KJ, Eberhardt H, Walz VM (1963) Arch Gewerbepathol Gewerbehyg 20: 41
58. Kuoni J (1980) Praxis 69: 1225
59. Martin G, Knorpp K, Huth K, et al (1968) Dtsch Med Wochenschr 93: 2002
60. Schönborn H, Prellwitz W, Baum P (1970) Klin Wochenschr 48: 822
61. Yokaiden RE, Babcock JR (1973) Arch Environ Health 26: 281
62. Dorndorf W, Kresse M, Christian W, Katritzki G (1975) Arch Psychiatr Nervenkrankh 220: 373
63. Bignami M, Cardamone G, Carere A, et al (1977) Mutat Res 46: 243
64. Brehm H, Stein AB, Rosenkranz HS (1974) Cancer Res 34: 2576
65. McCann J, Simmon V, Streitweiser D, Ames BN (1975) Proc Natl Acad Sci 72: 3190
66. McCann J, Spingarn NE, Lobori J, Ames BN (1975) Proc Natl Acad Sci 72: 979
67. Rannug U, Beije B (1979) Chem-Biol Interact 24: 265
68. Rannug U, Sundvall A, Ramel C (1978) Chem-Biol Interact 20: 1
69. Ehrenberg L, Osterman-Golkar S, Singh D, Lundqvist U (1974) Radiat Bot 15: 185
70. Theiss JC, Stoner GD, Shimkin MB, Weisburger EK (1977) Cancer Res 37: 2717
71. Ward JM (1980) Banbury Rep 5: 35
72. Maltoni C, Valgimigli L, Scarnatto C (1980) Banbury Rep 5: 5
73. Vozovaja MA (1976) Gig Sanit 6: 100
74. Vozovaya MA (1974) Gig i Sanit 7: 25
75. Rao KS, Murray IS, Deacon MM, et al (1980) Banbury Rep 5: 145
76. Reckner LR, Sachdev I (1975) DHEW (NIOSH) Publ No 75-184. U.S. Dept. of Health Education, and Welfare, Washington DC
77. Henschler D, Reichert Metzler M (1980) Int Arch Occup Environ Health 47: 263
78. National Institute for Occupational Safety and Health (NIOSH) (1977) NIOSH Manual of Analytical Methods. DHEW (NIOSH) Publ. No. 77-157B. U.S. Dept. Health, Education, and Welfare, Washington DC
79. Nicholson AA, Meresz O, Lemyk B (1977) Anal Chem 49: 814
80. Dowty BJ, Carlish DR, Laseter JL (1975) Environ Sci Technol 9: 762
81. Horwitz E (Ed.) (1975) Official Methods of Analysis of the Association of Official Analytical Chemists. 12th Ed, Assoc Off Anal Chem, Washington DC
82. Pearson C, McConnel G (1975) Proc R Soc, London Ser B 189: 305
83. Kretzschmar JG, Peperstraete H, Rymen T (1976) Extern 5: 147

84. PED Co Environmental, Inc. (1979) Monitoring of Ambient Levels of EDC near Production and User Facilities. EPA Rep. No 600/4-79-029. Environment Protection Agency (Off. Res. Develop.), Research Triangle Park NC
85. Kellam RG, Dusetzina MG (1980) Banbury Rep 5: 265
86. Suta B (1969): Assessment of Human Exposures to Atmospheric Ethylene Dichloride. SRI International, Menlo Park, CA
87. Okuno T, Tsuji M, Shintani Y, Watanabe H (1974) Chem Abstr 87: 72564
88. Jensen S. Lange R, Berge G, et al (1975) Proc R Soc London, Ser B 189: 333
89. Symons JM, Bellar TA, Carswell JK, et al. (1975) J Am Water Works Assoc 67: 634
90. Ewing BB, Chian ESK, Cook JC, et al (1977) EPA-560/6-77-015. U.S. Environ. Prot. Agency, Washington DC
91. Fuji T (1977) J Chromatogr 139: 297
92. Page BD, Kennedy BP (1975) J Assoc Off Anal Chem 58: 1062
93. Tute MS (1971) Adv Drug Res 6: 1
94. Aviado DM, Zakhari S, Simaan JA, Ulsamer AG (1976) Methyl Chloroform and Trichloroethylene in the Environment. Cleveland OH
95. International Agency for Research on Cancer (IARC) (1987) Overall evaluations of carcinogenicity: An updating of IRAC Monographs volume 1 to 42. IARC Monographs on the evaluation of carcinogenic risks to humans. Suppl. 7, WHO, Lyon
96. Stewart RD (1968) Ann Occup Hyg 11: 71
97. Hake CL, Waggoner TB, Robertson DN, Rowe VK (1960) Arch Environ Health 1: 101
98. VanDyke RA, Wineman CG (1971) Biochem. Pharmacol. 20: 463
99. Ikeda M, Ohtsuji H (1972) Br J Ind Med 29: 99
100. Riihimäki V, Pfäffli P (1978) Scand. J. Work. Environ. Health 4: 73
101. Morgan A, Black A, Belcher DR (1972) Ann. Occup. Hyg. 15: 273
102. Fernandez JG, Humbert BE. (1977) Arch Mal Prof 38: 415
103. Seki Y, Urashima Y, Aikawa H, et al (1975) Int Arch Arbeitsmed 34: 39
104. Gazzaniga G, Binaschi S, Sportelli A, Riva M (1969) Boll Soc Ital Biol Sper 45: 97
105. Monster AC (1979) Int Arch Occup Environ Health 42: 311
106. Monster AC, Houtkooper JM (1979) Int Arch Occup Environ Health 42: 319
107. Monster AC, Boersma G, Steenway H (1979) Int Arch Occup Environ Health 42: 293
108. Duprat P, Desault L, Grediski D (1976) Eur J Toxicol Environ Hyg 9: 171
109. Savolainen H, Pfäffli P, Tengen M, Vainio H (1977) Arch Toxicol 38: 229
110. Adams EM, Spencer HC, Rowe VK, Irish D (1950) Arch Ind Hyg Occup Med 1: 225
111. Klaassen CD, Plaa GL (1966) Toxicol Appl Pharmacol 9: 139
112. Torkelson TR, Oyen F, McCollister DD, Rowe VK (1958) Am Ind Hyg Assoc J 19: 353
113. McNutt NS, Amster RL, McConnell EE, Morris F (1975) Lab Invest 32: 642
114. Carlson GP (1973) Life Sci 13: 67
115. Klaassen CD, Plaa GL (1967) Toxicol Appl Pharmacol 10: 119
116. Lal H, ölshan A, Puri S, et al (1969) Toxicol Appl Pharmacol 14: 625A
117. Lal H, Shah H (1967) Toxicol Appl Pharmacol 10: 389A
118. VanDyke RA, Chenoweth MB (1965) Anesthesiology 26: 348
119. Priestley BG, Plaa GL (1976) Arch Int Pharmacodyn Ther 223: 132
120. Taylor GI, Drews RT, Lores EM, Clemmer TA (1976) Toxicol Appl Pharmacol 38: 379
121. Reinhardt CF, Mullin LS, Mayfield ME (1972) Toxicol Appl Pharmacol 22: 305
122. Rennick BR, Malton SD, Moe GK, Seevers MH (1949) Féd Proc Fed Am Soc Exp Biol 8: 327
123. Hermansen K (1970) Acta Pharmacol Toxicol 28: 17
124. Herd PA, Lipski M, Martin HF (1974) Arch Environ Health 28: 227
125. Plaa GL, Larson RF (1965) Toxicol Appl Pharmacol 7: 37
126. Gamberale F, Hultengren M (1973) Work Environ Health 10: 82
127. Salvini M, Binaschi S, Riva M (1971) Br J Ind Med 28: 286
128. Bass M (1970) Am Med Assoc J 212: 2075
129. Dornette WHL, Jones JP (1960) Anesthesia Analgesia 39: 249
130. Hall FB, Hine CH (1966) J Forensic Sci 11: 404
131. Hatfield R, Maykowski RT (1970) Arch Environ Health 20: 279
132. Kleinfeld M, Feiner B (1966) J Occup Med 8: 358
133. Stahl CJ, Fatteh AV, Dominguez AM (1969) J Forensic Sci 14: 393
134. Halevy J, Pitlik S, Rosenfeld J, Eitan BD, (1980) Clin Toxicol 16: 467
135. Texter EC, Grunow WA, Zimmerman HJ (1979) Clin Res 27: 684A

136. Kramer CG, Ott MG, Fulkerson JE, et al (1978) Arch Environ Health 33: 331
137. Simmon VF, Kauhanen K, Tardiff RG (1977) Mutagenic activity of chemicals identified in drinking water. In: Scott ID, Bridges BA, Sobels F (eds) Progress in Genetic toxicology. Amsterdam 249
138. Schwetz BA, Leong BK, Gehring P (1975) Toxicol Appl Pharmacol 32: 84
139. National Cancer Institute (NCI) (1977) Bioassay of 1,1,1-trichloroethane for Possible Carcinogenicity. DHEW Publ. No. (NIH) 77–803. U.S. Dept. Health, Education, and Welfare, Washington DC
140. Rampy LW, Quast IF, Leong BKI, Gehring PI (1978) Results of long term inhalation toxicity studies on rats of 1,1,1-trichloroethane and perchloroethylene formulations. In: Plaa GL, Duncan WAM (eds.): Proceedings of the First International Congress of Toxicology, New York
141. Elovaara E, Hemminki K, Vainio H (1978) Toxicology 12: 111
142. Russel JW, Shadoff LA (1977) J Chromatogr 134: 375
143. Lovelock JE (1977) Ecotoxico Environ Saf 1: 399
144. McConnell G, Ferguson DM, Pearson CR (1975) Endeavour 34: 13
145. Spence JW, Houst PL (1978) J Air Pollut Control Assoc 28: 250
146. Cox RA, Derwent GR, Eggleton AEJ, Lovelock JE (1976) Atmos Environ 10: 305
147. Grimsrud EP, Rasmussen RA (1975) Atmos Environ 9: 1014
148. Lillian D, Singh HB, Appleby A, et al (1975) Environ Sci Technol 9: 1042
149. Ohta T, Morita M, Miziguchi J (1976) Atmos Environ 10: 557
150. Singh HB, Salas LI, Cavanaugh LA (1977) J Air Pollut Control Assoc 27: 332
151. Correira Y, Martens GJ, van Mensch FH, Whim BP (1977) Atmos Environ 11: 1113
152. Neumayer V (1981) WaBoLu-Berichte 3: 24
153. Dahlberg A, Christiansen VO, Eriksson EA (1973) Ann Occup Hyg 16: 41
154. Yllner S (1971) Acta Pharmacol Toxicol 29: 499
155. Wahlberg JE (1976) Ann Occup Hyg 19: 115
156. Gehring PJ (1968) Toxicol Appl Pharmacol 13: 287
157. Hanasono GK, Witschi H, Plaa GL (1975) Proc Soc Exp Biol Med 149: 903
158. Kluwe WM, Herrmann CL, Hook JB (1979) J Toxicol Environ Health 5: 605
159. Rosenberg R, Grahn O, Johansson L (1975) Water Res 9: 607
160. Chierutini ME, Franklin CS (1976) Br J Pharmacol 57: 421
161. Prost G, Rigaud M, Pelletier N (1977) Arch Mal Prof 38: 205
162. Truhaut R, Lich NP, Dutertre-Catella H, et al (1974) Arch Mal Prof 35: 593
163. Truhaut R, Thevenin M, Warnet M (1975) Eur J Toxicol Environ Hyg 8: 175
164. Yllner S (1971) Acta Pharmacol Toxicol 29: 481
165. Gohlke R, Schmidt P, Bahmann H (1977) Z Ges Hyg, Ihre Grenzgebiete 23: 278
166. Vainio H, Parkki MG, Marniemi J (1976) Xenobiotica 6: 599
167. Schmidt P, Burck D, Buerger A (1980) Z Gesamte Hyg, Ihre Grenzgeb 26: 167
168. Wolff DL, Siegmund R (1978) Biol Zentralbl 97: 345
169. Gobbato F, Slavich G (1968) Med Lav 59: 667
170. Lobo-Mendoca R (1963) Br J Ind Med 20: 50
171. Zollinger F (1931) Arch Gewerbepathol Gewerbehyg 2: 298
172. Callen DF, Wolf CR, Philpot RM (1980) Mutat Res 77: 55
173. Schmidt R (1976) Biol Rundschau 14: 220
174. Shackleford WM, Keith LH (1976) Frequency of Organic Compounds Identified in Water. EPA-600/4-76-062. U.S. Environ. Prot. Agency, Atlanta GA
175. Barsoum GS, Saad K (1934) Quart J Pharm Pharmacol 7: 205
176. Lehmann KB, Flury F (1943) Toxicology and Hygiene of Industrial Solvents. Translated by E King and HF Smyth Jr. Williams and Wilkins C, Baltimore ML
177. Joachimoglu G (1921) Berl Klin Wochenschr 58: 147
178. Fowler JSK (1969) Br J Pharmacol 35: 530
179. Jondorf WR, Parke DV, Williams RT (1957) Biochem J 65: 14
180. Bray HG, Thorpe WV, Vallance DK (1952) Biochem J 51: 183
181. Weeks MH, Angerhofer RA, Bishop R, et al. (1979) Am Ind Hyg Assoc J 3: 187
182. Keith LH (1976) Environ Sci Technol 10: 555
183. Keith LH (ed) (1976) Identification and Analysis of Organic Pollutants in Water. Ann Arbor Sci Publ, Ann Arbor

Organic Explosives and Related Compounds

David H. Rosenblatt, Elizabeth P. Burrows, Wayne R. Mitchell, and David L. Parmer

U.S. Army Biomedical Research and Development Laboratory, Fort Detrick, Frederick/Maryland 21701-5010, USA

Summary

For much of this century production and usage of explosives and propellants have been responsible for release to the environment of a variety of energetic organic nitro compounds. This chapter covers the compounds of greatest importance; their uses are indicated and methods for their manufacture or laboratory synthesis are summarized. Those physicochemical properties of greatest utility in environmental risk assessment are identified and listed. Analytical methods are described briefly, with references given to more detailed literature. For the more extensively studied compounds, the microbiological and chemical transformations known to take place in the laboratory and in the environment and the metabolic transformations observed in animals and man are discussed. Toxic manifestations in mammals, humans, fish and other aquatic organisms, as well as threshold levels for these effects, are summarized. While the process of utilization of available experimental data to develop criteria and standards to assure protection of human health and preservation of the biosphere is in its infancy and changing rapidly, a short summary of criteria currently accepted in the U.S. for certain of the munitions compounds is presented.

The Handbook of Environmental Chemistry,
Volume 3 Part G, Ed. O. Hutzinger
© Springer-Verlag Berlin Heidelberg 1991

Introduction

Explosives and propellants have important military applications; the former are also widely used in mining and construction. Their manufacture represents a sizable segment of the chemical industry [1]. In the course of production, handling, loading of military or civilian devices, and ultimate dispersal or disposal, explosives and propellants are released to the environment. There they are disseminated by natural processes and partially converted to secondary products. This chapter deals only with the more important organic explosives and propellants, and the focus is primarily on physicochemical properties and behavior, environmental fate, toxicity to human beings and wildlife, and environmental criteria; a comprehensive review that emphasizes manufacturing processes, formulations and uses is available [2]. While reasonable attempts have been made to assemble all pertinent material, some sources will doubtless have been missed. In addition to such omissions, there are substantial gaps in our knowledge, especially of environmental fate and chronic toxicity. Table 1 is a summary of the compounds to be covered and the abbreviated names for them that will be used throughout the chapter.

Table 1. Listing of explosives, propellants, and derived substances

Compound[a]	Abbreviation
Trinitrotoluene[b]	TNT
Dinitrotoluene[c]	DNT
1,3,5-Trinitrobenzene	TNB
1,3-Dinitrobenzene	DNB
Hexahydro-1,3,5-trinitro-1,3,5-triazine	RDX
Octahydro-1,3,5,7-tetranitro-1,3,5,7-tetrazocine	HMX
1-Acetylhexahydro-3,5-dinitro-1,3,5-triazine	TAX
1-Acetyloctahydro-3,5,7-trinitro-1,3,5,7-tetrazocine	SEX
N,2,4,6-Tetranitro-N-methylaniline	Tetryl
Ammonium picrate/Picric acid	AP/PA
Pentaerythritol tetranitrate	PETN
Nitroglycerin (glyceryl trinitrate)	NG
Nitroguanidine	NQ
Ethylene glycol dinitrate	EGDN
Diethylene glycol dinitrate	DEGDN
Propylene glycol dinitrate	PGDN

[a] Structures are presented in Fig. 1.

[b] Where TNT is used, the 2,4,6-isomer (formerly known as α-TNT) is denoted; other isomers are specifically designated, e.g., 2,3,4-TNT.

[c] Where DNT is used, the 2,4-isomer is denoted, possibly with minor amounts of other isomers – especially the 2,6-isomer; other isomers are specifically designated, e.g., 2,3-DNT.

Synthesis/Production and Use

Methods for the synthesis or industrial production and purification of the compounds of interest are summarized and referenced in Table 2. TNT, historically the most important and most widely used military high (or "secondary") explosive, has not been produced in the U.S. for several years. It is currently an important constituent, along with RDX (see below), of formulations known as Composition B. For the most part, the other five isomers present in crude TNT were converted by the "sellite" (i.e., sodium sulfite) process for TNT purification [3] to water-soluble dinitrotoluenesulfonate salts; the latter were destroyed by incineration of the "red water" concentrates in which they occurred. Thus, they are ordinarily not observed as environmental pollutants. The dinitrotoluenes were present in TNT wastewaters and in the purified product (military-grade TNT); 2,4-DNT is listed as a priority pollutant, and both 2,4- and 2,6-DNT have been detected as soil and groundwater contaminants. DNT is also utilized in its own right as a propellant constituent to control burning rates or to reduce hygroscopicity [4]. TNB is a stable end-product of TNT photolysis that occurs wherever TNT has been a soil surface contaminant or a water contaminant over a period of time [5]. DNB apparently arose through the nitration of benzene, present as an impurity of toluene, during the manufacture of TNT.

The most important military high explosive in the U.S. today is RDX (British code name for *Research Department* or *Royal Demolition* Explosive). It

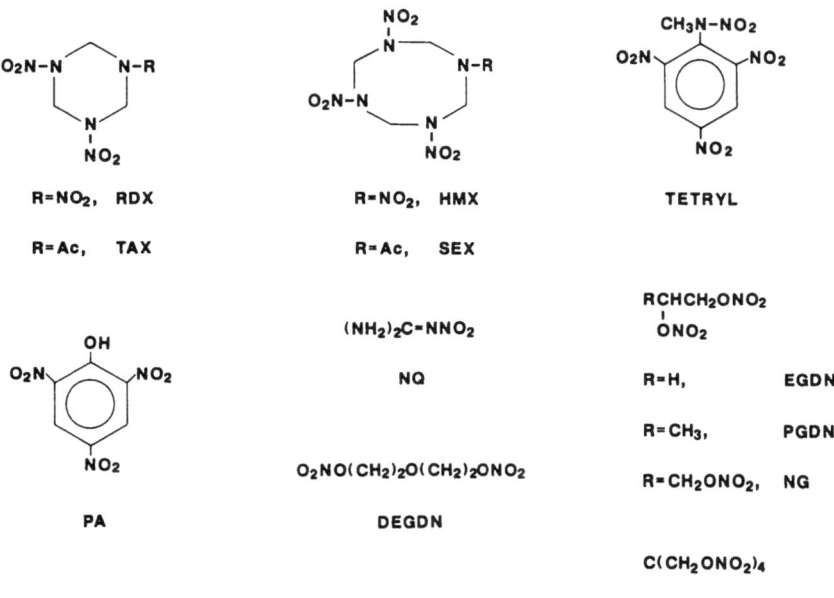

Fig. 1. Structures of Explosives and Related Compounds

Table 2. Synthesis/production processes and purification of explosives, propellants, and related compounds

Compound[a]	Synthetic Process	Purification	References
TNT[b]	Nitration of toluene	Isomer removal with aqueous sodium sulfite	[3, 7]
DNT[c]	Nitration of toluene	"Sweating[d]"	[4, 8]
2,4-DNT	Nitration of 4-nitro-toluene	Recrystallization	[8]
2,6-DNT	Nitration of 2-nitro-toluene	Distillation	[4, 8]
TNB	Decarboxylation of 2,4,6-trinitrobenzoic acid	Recrystallization	[9]
DNB	Nitration of benzene	Isomer removal with aqueous sodium sulfite	[10]
RDX	Nitrolysis of hexa-methylenetetramine	Recrystallization	[11, 12]
HMX	Nitrolysis of hexa-methylenetetramine	Recrystallization	[13, 14]
TAX	Nitrolysis of 1,3,5-triacetylhexahydro-1,3,5-triazine	Chromatography plus recrystallization	[15]
SEX	Nitrolysis of 1,5-diacetyloctahydro-3,7-dinitro-1,3,5,7-tetrazocine	Chromatography plus recrystallization	[16]
Tetryl	Nitration of N,N-dimethylaniline	Recrystallization	[17]
NQ	Dehydration of guanidine nitrate	Recrystallization	[6]
PA/AP[e]	Sulfonation, then nitration of phenol; ammoniation	Water washing	[18]
PETN	Nitration of pentaerythritol	Recrystallization	[19]
NG[f]	Nitration of glycerol	Washing, drying	[20]
EGDN	Nitration of ethylene glycol	Probably washing	[21]
DEGDN	Nitration of diethylene glycol	Washing	[22]
PGDN	Nitration of propylene glycol	Probably washing	[23]

[a] Italicized compounds are the munitions compounds of primary interest; others are impurities or byproducts associated with the previously listed italicized compounds.

[b] For syntheses of the isomers of 2,4,6-trinitrotoluene see Refs. [24 and 25].

[c] For syntheses of the 2,3-, 2,5-, 3,4- and 3,5- isomers of DNT, see Refs. [8, 24 and 26].

[d] "Sweating" of DNT [27] involves the gradual warming in a sieve-like "sweat pan" of a crystalline mixture of DNTs containing mostly the 2,4-isomer. The 2,6-isomer, along with minor constituents, forms a low-melting eutectic fraction that drains from the crystal mass through the holes in the pan. The melting point of the residual material in the pan is thereby raised, as is the content of 2,4-DNT.

[e] Picramic acid, a metabolic product of picric acid, is prepared by the reaction of picric acid with sodium sulfide [28].

[f] For syntheses of the mono- and dinitrate esters of glycerol, see Refs. [29, 30, 31 and 32].

is formulated into munitions alone (Composition A) or as a slurry with molten TNT (Composition B). Production grade RDX generally contains significant amounts of HMX (code name for High Melting Explosive) as an acceptable impurity. By modification of the RDX synthetic process, HMX becomes the major product, with RDX as an impurity. HMX is being increasingly used, predominantly as a propellant and also in maximum-performance explosives. TAX and SEX are both common impurities and wastewater constituents associated with RDX/HMX manufacture.

Tetryl was formerly produced mainly as a booster explosive. It has not been produced in over a decade and has been superseded in this role by RDX.

Picric acid (PA) and especially its salt, ammonium picrate (AP), are reputedly rather insensitive to shock; they have been used as burster charges in naval projectiles.

PETN, employed in blasting caps and detonators, is easily initiated and exhibits highly reproducible responses.

Pure nitroglycerin (NG) is too sensitive and unstable for a practical high explosive. As an energetic plasticizer, the compound is added to nitrocellulose in blasting gelatins and is a component of most double-base and triple-base propellants. Glyceryl mono- and dinitrates (MNGs and DNGs) are found in the washwaters of nitroglycerin manufacturing plants.

Nitroguanidine (NQ) is included in triple-base propellants to impart flashlessness and to reduce erosivity [6].

EGDN is a liquid propellant component used in land mines to eject small projectiles upward to bursting heights.

DEGDN is more stable and less shock-sensitive than NG. It is replacing NG as a propellant plasticizer for some munition items.

PGDN is a component of "Otto fuel", a liquid naval torpedo propellant composed of 76% PGDN, 22.5% di-n-butyl sebacate, and 1.5% 2-nitrodiphenylamine [33].

Properties and Estimation Methods

In this review, the only properties presented are those likely to be of value in making environmental calculations and decisions. For such use-related characteristics as detonation velocity and thermodynamics, the reader is referred to Lindner's article in the Encyclopedia of Chemical Technology [2].

The vapor pressure of a compound is commonly given for the state, solid or liquid, in which the substance is likely to occur under ambient conditions, typically 25 °C. If the vapor pressure has been evaluated for a range of temperatures, the data may be presented as the constants of the Clausius–Clapeyron equation, $\log P = A - B/T$, where T is in degrees kelvin (K). Table 3 gives values for A and B (pressure in torr) for all but three (i.e., AP, TAX, and SEX) of the compounds of interest. For a compound that is solid at ambient temperature, calculation of the Henry's Law constant, K_H, may require that the vapor pressure of the supercooled liquid be converted to the vapor pressure of the solid. The logarithm of the required conversion factor may be calculated by the equation:

$$\log_{10}\{P_{(s)}/P_{(l)}\} = (\Delta S_f/2.303R)(T - T_m)/T, \tag{1}$$

where $R = 1.987$ cal/mol K, ΔS_f (entropy of fusion) is in the same units, and T is in degrees K [34]. When ΔS_f is not known, a value of 13.5 cal/mol K may be assumed for rigid molecules (whose definition for the present purposes includes all the solids discussed here) [35]. For the calculation at 25 °C (298.16 K), the equation reduces to:

$$\log_{10}\{P_{(s)}/P_{(l)}\} = 0.00989[25 - T_m(°C)]. \tag{2}$$

When the heat of fusion, ΔH_m (after conversion into cal/mol), is known in addition to the melting point (T_m), one should use the relationship $\Delta S_f = \Delta H_m/T_m$ to obtain the entropy of fusion; this relationship depends on the fact that the free energies of the solid and liquid phases are identical at the melting point, so that ΔF_f (the free energy change in going from the solid to the liquid state) at that temperature is zero.

Table 3. Constants for the Clausius–Clapeyron vapor pressure equation[a]

Compound	A[b]	B	Reference
TNT	14.53; (12.31 ± 0.34)	5900; (5175 ± 105)	[43; (44)]
2,4-DNT	13.08 ± 0.19	4992 ± 59	[44]
2,6-DNT	13.99 ± 0.18	5139 ± 52	[44]
DNB[c]	8.56	3170	[45]
TNB	13.29	5608	[43]
RDX	15.12; (11.87)	7011; (5850)	[43; (46)]
HMX	14.72	8407	[43]
Tetryl	15.19	6987	[43]
PA	12.28	5488	[43]
PETN	18.08	7856	[43]
NG[d]	9.20 ± 1.30	3525 ± 399	[47]
NQ	14.15	7452	[43]
EGDN	10.55 ± 0.08; (11.69 ± 0.52)[d]	3476 ± 22; (3830 ± 159)[d]	[44; (47)]
DEGDN[d]	9.97 ± 0.56	3637 ± 17	[47]
PGDN[d]	10.19 ± 0.30	3338 ± 93	[47]

[a] Log_{10} P (torr) $= A - B/T$ (K). Indicated uncertainties are standard errors.
[b] Values of A in reference 43 were converted from units of log Nm^{-2} (Newtons per square meter, or Pascal) to torr by subtracting 2.125 (i.e., log 133.3).
[c] The equation for DNB is for the supercooled liquid. The value of the vapor pressure at 25 °C would be obtained by use of the liquid-to-solid state conversion factor presented in the text.
[d] Values of A and B were derived from vapor pressures presented in the reference for 15, 25, 35, 45 and 55 ° C.

Solubilities of the munitions compounds at 25 °C are given in Table 4; some of these values represent limited interpolations from those found in the literature. Aqueous solubilities at temperatures other than 25 °C have been recorded for many of the substances of interest by Lindner [2], Urbanski [6,17–19,36,37], Fedoroff and Sheffield [38,39], Kaye [4], Roth [40], and Spanggord et al. [5]. Where no solubility is available, it may be estimated from other properties, such as the octanol-water partition coefficient (K_{ow}), or from structural information [41]. Banerjee et al. [42] demonstrated, using RDX as an example, that the correlation of solubility with K_{ow} is considerably improved if the entropy of fusion is taken into account; this compensates for the fact that high-melting compounds are generally less soluble than liquids or low-melting compounds of otherwise similar properties. Their equation (applicable at 25 °C) treats liquids as if they melted at 25 °C:

$$\log K_{ow} = 6.5 - 0.89 \log S \; (\mu mol/L) - 0.015 \; t_m(°C) \tag{3}$$

or, rearranged into:

$$\log S \; (\mu mol/L) = 7.30 - 1.12 \log K_{ow} - 0.01685 \; t_m \; (°C) \tag{4}$$

Table 4. Properties of explosives, propellants and related compounds

Property	Compound			
	TNT	2,4-DNT	2,6-DNT	TNB
CAS Reg. No. [63]	118-96-7	121-14-2	606-20-2	99-35-4
Empirical Formula	$C_7H_5N_3O_6$	$C_7H_6N_2O_4$	$C_7H_6N_2O_4$	$C_6H_3N_3O_6$
Molecular Weight	227.15	182.15	182.15	213.12
Density (g/cm³)	1.65 [2]	1.521 [8]	1.538 [8]	1.63 [36]
Melting Point (°C)	80.75 [2]	72 [64]	66 [64]	122 [9]
Heat of Fusion (cal/g)	23.5 [2]	26.1 [8]	22.5 [8]	16.0 [65]
Vapor Pressure (torr, 25° C)	5.51×10^{-6a} $(8.02 \times 10^{-6}$ [44]) $(3.7 \times 10^{-6})^b$	2.17×10^{-4a}	5.67×10^{-4a}	3.03×10^{-6a}
Aqueous Solubility (mg/L, 25 °C)	150 [37]	280^c	208 [50]	$\sim 385^d$
K_H^e (atm·m³/mole,25 °C)	1.10×10^{-8}	1.86×10^{-7}	$\sim 4.86 \times 10^{-7}$	2.21×10^{-9}
Diffusion Coefficient^f (air) (cm²/s)	0.064	0.067	0.067	0.068
Diffusion Coefficient^g (water) (cm²/s)	6.71×10^{-6}	7.31×10^{-6}	7.31×10^{-6}	7.20×10^{-6}
Log K_{ow}	2.00 [50]; 1.86 [51]; (1.84)^h	1.98 [50,52]	1.89 [50]; 2.02 [51]	1.18 [52]
Log K_{oc}	2.72 [66]	2.40 [66]	1.89^i	1.30^i
BCF (fish)	8.95^j	10.6 [57]; 11.6^j; 3.8^k	9.82^j	2.65^j
BCF^l (fat/feed, beef)	0.0029	0.0034	0.0031	0.0014

Table 4. (*Continued*)

Property	Compound			
	DNB	RDX	HMX	Tetryl
CAS Reg. No. [63]	99-65-0	121-82-4	2691-41-0	479-45-8
Empirical Formula	$C_6H_4N_2O_4$	$C_3H_6N_6O_6$	$C_4H_8N_8O_8$	$C_7H_5N_5O_8$
Molecular Weight	168.12	222.15	296.20	287.17
Density (g/cm^3)	1.575 [48]	1.83 [2]	1.90 (β form) [2]	1.73 [2]
Melting Point (°C)	90 [48]	205 [42]	286 [2]	129.5 [2]
Heat of Fusion (cal/g)	28.97 [67]	38.26 [68] 32.86 [69]	m	21.6 [68]
Vapor Pressure^a (torr, 25 °C)	1.93×10^{-4}	4.03×10^{-9}	3.33×10^{-14}	5.69×10^{-9}
Aqueous Solubility (mg/L, 25 °C)	533 [70]	60 [42]	5 [71]	80 [17]
K_H^e (atm·m^3/mole, 25 °C)	8.01×10^{-7}	1.96×10^{-11}	2.60×10^{-15}	2.69×10^{-11}
Diffusion Coefficient^f (air) (cm^2/s)	0.073	0.074	0.063	0.059
Diffusion Coefficient^g (water) (cm^2/s)	7.94×10^{-6}	7.15×10^{-6}	6.02×10^{-6}	5.99×10^{-6}
Log K_{ow}	1.49 [52]	0.87 [42]; 0.81 [50]; 0.86 [51]	0.26 [50]; 0.06 [51]	1.65 [51]
Log K_{oc}	1.56^i	2.00 [66]	0.54^i	1.69^i
BCF^j (fish)	4.70	1.50	0.49	6.31
BCF^l (fat/feed, beef)	0.0019	0.00095	0.00047	0.0023

The Henry's Law constant, K_H (or H), is often expressed in terms of (atmospheres · cubic meters)/mol; it is a measure of the ratio of the concentration in the gaseous state to the concentration in solution (here restricted to aqueous solution). In the case of a substance that is appreciably dissociated (or associated) to ionic species, only the concentration of the uncharged species is relevant. Thus, PA is highly dissociated, with a pK_a of 0.38 [48], so that only 0.000025% remains as the undissociated molecule at pH 7; essentially, then, dilute picric acid in groundwater should have no vapor pressure. For Table 4, the following equation was used to calculate K_H from the vapor pressure (torr or mm of mercury), molecular weight, and solubility (mg/L):

$$K_H \text{ (atm · m}^3/\text{mol)} = (P \times MW)/(S \times 760) \tag{5}$$

If K_H is expressed in torr M^{-1}, it may be converted to (atmospheres · cubic meters)/mole through division by 760 000; K_H at 25 °C in (atmospheres · cubic meters)/mole can be converted to the dimensionless form through multiplication by the constant 40.88 mol/(m^3 · atm).

Octanol-water partition coefficients are usually given in the logarithmic form, log K_{ow} (or log P_{ow}). K_{ow} is defined as the ratio of a dissolved chemical's concentration in octanol (saturated with water) to its concentration in water

Table 4. (*Continued*)

Property	Compound			
	PA	NQ	PETN	NG
CAS Reg. No. [63]	88-89-1	556-88-7	78-11-5	55-63-0
Empirical Formula	$C_6H_3N_3O_7$	$CH_4N_4O_2$	$C_5H_8N_4O_1$	$C_3H_5N_3O_9$
Molecular Weight	229.12	104.09	316.17	227.11
Density (g/cm³)	1.76 [2]	1.72 [2]	1.78 [2]	1.59 [2]
Melting Point (°C)	123 [2]	245 (dec) [2]	141 [2]	13.2 [2]
Heat of Fusion (cal/g)	20 [2]	m	76 [2]	m
Vapor Pressure (torr, 25 °C)	7.47×10^{-7a}	1.43×10^{-11a}	5.38×10^{-9a}	0.00177 [47]
Aqueous Solubility (mg/L, 25 °C)	12,400[n]	4200 [39]	2.1 [72]	1950 [73]
K_H^e (atm·m³/mole, 25 °C)	o	4.67×10^{-16}	1.07×10^{-9}	2.71×10^{-7}
Diffusion Coefficient[f] (air) (cm²/s)	0.066	0.102	0.057	0.070
Diffusion Coefficient[g] (water) (cm²/s)	7.03×10^{-6}	1.04×10^{-5}	5.61×10^{-6}	6.95×10^{-6}
Log K_{ow}	2.03 [52]	− 0.83 [74];	(3.71)[p]	1.62 [75]; 1.77 [51] (2.81)[p]
Log K_{oc}	2.00[i]	− 0.356[i]	3.39[i]	2.77 [76]; 1.66[i]
BCF (fish)	12.7[j,q]	0.065[j]	281[j]	≤ 15 [77]
BCF[l] (fat/feed, beef)	0.0036	0.00013	0.025	0.0023

(saturated with octanol) when the two liquid phases are at equilibrium, i.e., $K_{ow} = C_o/C_w$. For environmental purposes, K_{ow} is not employed in its own right, but is frequently the starting point for estimating directly useful properties such as solubility, bioconcentration factors or soil organic carbon/water partition coefficient (K_{oc}) [49].

Experimental values for log K_{ow} of TNT determined independently in our laboratory [50] and another [51] are given in Table 4, and are in good agreement with the value calculated from experimental log K_{ow} values of 2,4-DNT or TNB through fragment substitution [52]. Experimental values have been published for three of the five nitrate esters considered here, NG, EDGN, and DEGDN (see Table 4). Values of log K_{ow} calculated for the five compounds using the π approach [52] are also included parenthetically, but are used for the estimations described below only in the cases of PETN and PGDN, where experimental data are lacking.

The soil organic carbon/water partition coefficient, K_{oc}, is defined as (μg adsorbed chemical per g organic carbon)/(μg chemical per mL of solution) [53]. Estimation of the actual partition coefficient, K_d, of a chemical between soil and water is based on the assumption that the soil's organic content is the only determinant of the sorption of a compound from water to soil. Thus, $K_d = f_{oc} \times K_{oc}$, where f_{oc} is the fraction of organic carbon in the soil. The organic

Table 4. (*Continued*)

Property	Compound		
	EGDN	DEGDN	PGDN
CAS Reg. No. [63]	628-96-6	693-21-0	6423-43-4
Empirical Formula	$C_2H_4N_2O_6$	$C_4H_8N_2O_7$	$C_3H_6N_2O_6$
Molecular Weight	152.08	196.14	166.11
Density (g/cm^3)	1.49 [2]	1.38 [2]	1.37 [3]
Melting Point (°C)	-22.8 [2]	-11.3 [2]	< -20 [23]
Heat of Fusion (cal/g)	m	m	m
Vapor Pressure (torr, 25 °C)	0.0706 [47]	0.00593 [47]	0.0984 [47]
Aqueous Solubility (mg/L, 25 °C)	5600 [38]	4000 [2]	3500[r]
K_H^e (atm·m^3/mole, 25 °C)	2.52×10^{-6}	3.83×10^{-7}	6.14×10^{-6}
Diffusion Coefficient[f] (air) (cm^2/s)	0.084	0.069	0.077
Diffusion Coefficient[g] (water) (cm^2/s)	8.72×10^{-6}	7.05×10^{-6}	7.93×10^{-6}
Log K_{ow}	1.16 [75]; (2.11)[p]	0.98 [74]; (1.13)[p]	(2.66)[p]
Log K_{oc}	1.28[i]	2.03 [74]	2.52[i]
BCF[j] (fish)	2.6	1.84	40.6
BCF[l] (fat/feed, beef)	0.0013	0.0011	0.0075

[a] Vapor pressures were calculated from constants in Table 3 by use of the Clausius-Clapeyron equation. Note the correction required for DNB (footnote d to Table 3).

[b] The vapor pressure for TNT at 25° C was estimated by linear interpolation between data for 21.5 and 25.5° [78].

[c] The solubility of 2,4-DNT was estimated by linear interpolation between data for 22° and 50 °C [4].

[d] The solubility of TNB was estimated by interpolation of a plot of ln S [36] vs. 1/T (K) for 17°, 50°, and 100 °C (i.e., 290, 323, and 373K).

[e] K_H values were calculated from listed values of vapor pressure and solubility [Eq. (5)].

[f] Diffusion coefficients in air were estimated according to the FSG method described in chapter 17 of Ref. [49].

[g] Diffusion coefficients in water were estimated according to the Hayduk-Laudie method (chapter 17 of Ref. [49]), which requires judgement in the choice of some of the volume increments. A value of 20 for the 8-membered ring was used for HMX.

[h] This value was calculated from log K_{ow} values for 2,4-DNT or TNB through fragment substitution [52].

[i] Values of log K_{oc} were calculated by the equation of Lyman and Loreti [54] [Eq (6)].

[j] Values of BCF (fish) were calculated from log K_{ow} by the equation of Isnard and Lambert [56] [Eq. (7)].

[k] This value was estimated by USEPA [58].

[l] Values for BCF (fat/feed, beef) were estimated from log K_{ow} by the equation of Kenaga [59] [Eq. (8)].

[m] Information not available

[n] The solubility of PA at 25° C was estimated by linear interpolation between values for 20° and 30° C [18].

[o] K_H was not calculated for PA, as explained in the text.

[p] This value was calculated from π constants [52] (see text).

[q] A value of < 1 in trout epaxial muscle was found experimentally [79] (see Metabolism section).

[r] The solubility of PGDN was estimated from log K_{ow} according to Eq. (4) [42].

matter content may be converted to f_{oc} by multiplying it by 0.58 [53]. A number of equations have been developed to derive log K_{oc} from other environmentally relevant partition coefficients [53, 54]. An example, which has been applied to a wide variety of organic solutes, is that of Lyman and Loreti [54]:

$$\log K_{oc} = 0.824 \log K_{ow} + 0.328. \tag{6}$$

BCF (fish), the water-to-aquatic organism (or simply "fish") bioconcentration factor, is defined as a concentration ratio determined at equilibrium (see Ref. [55]), namely, chemical concentration in flesh (wet weight)/chemical concentration in water. In practice, many published determinations were made under non-equilibrium conditions. Where experimental data or literature values were unavailable and estimates of BCF (fish) values were required for the present purposes, they were calculated from the octanol-water partition coefficients [56]:

$$\log BCF = 0.80 \log K_{ow} - 0.52. \tag{7}$$

For 2,4-DNT, an experimental value in bluegill striated muscle [57] was in good agreement with that calculated, and exceeded that estimated by the U.S. Environmental Protection Agency (EPA) [58] for edible fish flesh.

An equation for the bioconcentration factor, BCF (fat/feed, beef), based on the partitioning of chemicals between animal feed and the fat of the beef, was published by Kenaga in 1980 [59] and has been used for estimations in the present work (Table 4):

$$\log BCF = -3.457 + 0.500 \log K_{ow}. \tag{8}$$

In the same paper, Kenaga gave regressions based on water solubility, K_{oc}, and fish BCFs; he also presented regressions for fish BCFs and swine BCFs. These and other regression models were compared, evaluated, and extended to other mammals and poultry by Garten and Trabalka [60]. Most recently, Travis and Arms [61] have presented equations to relate log K_{ow} to "biotransfer factors" for organic chemicals in cattle (flesh and milk). They also presented a regression for vegetation bioconcentration based on log K_{ow}.

While such equations have some usefulness in predicting bioaccumulation potential, substantial underestimations, especially for the less lipophilic compounds, may be expected if mechanisms unrelated to fat/water partitioning are oimportant. As an example, for this report we utilized bioaccumulation data for RDX in rats (see Metabolism section) and assumed a fat (lipid) content of 10% [62] in the rat to estimate a BCF (fat/feed) of 0.14, more than 100 times the Kenaga value for BCF (fat/feed, beef) (see Table 4). Such a comparison obviously ignores effects due to interspecies differences, but suggests that physicochemical regression models alone do not necessarily provide reliable estimates of bioaccumulation.

Analytical Methods

A comprehensive review of this topic was published by Yinon in 1977 [80]. Therefore this section will highlight some of the more recent contributions and trends in the field.

For more than a decade high-pressure liquid chromatography (HPLC) has been the most versatile and most widely used method for separating mixtures of munitions compounds. It is applicable to all classes of organic munitions and is especially useful for the thermally unstable nitramines and nitrate esters. A number of reversed-phase sorbents are available, and, with water-methanol or water-acetonitrile as the mobile phase, aqueous samples can be injected directly. While detection and quantitation by UV absorbance is simple and most frequently used [81–88], methods specific for detection of nitro groups such as electrochemical oxidation-reduction [89–93] and thermal energy analysis (TEA) [94, 95] have also been reported. For positive identifications of compounds present in low nanogram to picogram quantities the HPLC column is coupled to a mass spectrometer (MS) operating in positive [96] or negative [97, 98] chemical ionization mode (PCI or NCI) with the HPLC mobile phase (normally methanol-water) serving as the ionizing gas. Munitions compounds well characterized by this process (thermospray HPLC/MS) include TNT, DNT, RDX, HMX, NG, PA, PETN, and DEGDN.

Gas chromatography (GC) with electron capture (EC) detection continues to be used for separation and quantitative analysis of mixtures containing the more volatile, thermally stable compounds [99–102]. In a recent study, TEA was compared with EC for GC detection and was found preferable at low ng levels [103]. The use of macroreticular resins as an alternative to solvent extraction to remove and concentrate munitions compounds from groundwater samples prior to GC/EC analysis [104] or to HPLC analysis [92] has also been described. The earliest method coupling separation with profiles unique to each constituent of the mixture, GC/MS, is still commonly used for positive identifications of munitions compounds amenable to GC, and use of both electron impact (EI) MS [105] and CIMS [105, 106] modes gives complementary information. The MS characteristics and fragmentation pathways observed for TNT, DNT, RDX, HMX, NG, PA, PETN, and tetryl in PCI and NCI modes with a variety of ionizing gases (methane, isobutane, and ammonia) have been recorded and are amply documented [107–109]. Field desorption (FD) MS of RDX, HMX, TNT, PETN, and NG have also been reported [110].

Tandem mass spectrometry (MS/MS) [111] constitutes a new and powerful technique for the separation and characterization of mixtures. Picogram amounts of a desired analyte in a complex mixture can frequently be determined without prior GC or LC separation. The molecular ion or other prominent ion characteristic of the analyte is selected and its fragmentation pathway can be verified, without interference from other ions due to other components of the mixture. An investigation of the fragmentation pathways of TNT by high resolution EI MS/MS using ^{15}N and ^2H labeling revealed that the predominant processes are loss of OH followed by loss of NO or NO_2 [112]. These pathways were confirmed and studied in detail for a series of ten 2,4,6-trinitroaromatics [113]. A study of the high resolution MS/MS fragmentation pathways of RDX and HMX in EI, PCI, and NCI modes showed that a large number of fragment ions originate from formation of adduct ions $[M + NO]^+$ and $[M + NO_2]^+$ in EI and PCI, and from $[M + NO]^-$ and $[M + NO_2]^-$ in NCI, followed by dissociation [114]. Additional MS/MS measurements in EI, PCI, and NCI modes reported for RDX, HMX, TNT, DNT, PETN, NG, EGDN, and PA demonstrated that, for a given ionization mode, compounds in the same class (i.e., nitroaromatics, nitramines, and nitrate esters) exhibit losses of the same fragments [115]. Measurements of NCI MS/MS for TNT, DNT, 2,6-DNT, and RDX have also been made with an ion trap detector [116]. A review of the uses of mass spectrometry for detection and analysis of explosives in forensic chemistry has also been published recently [117].

Numerous TLC protocols for munitions compounds have been published in the 1960s and 1970s [80], but the method is of little more than historical importance today. It is both less sensitive and more cumbersome and time-consuming than HPLC or GC. Nevertheless, a protocol for two-dimensional TLC separations of multiclass components of explosives [118] and an improved procedure for nitrate esters and their metabolites and/or hydrolysis products [119] have appeared recently.

Environmental Fate

Environmental Entry and Distribution

The major route of environmental entry for munitions compounds is the discharge of waste streams generated during their manufacture and processing to surface water. The levels in such wastewaters vary widely, depending on the intensity of the manufacturing operations and efficacy of the treatment technologies employed. For example, numerous surveys of effluent TNT levels at munitions plants have reported concentrations ranging from < 0.05 mg/L to near saturation in the worst cases [120]. On the other hand, effluents at Holston Army Amunition Plant, Kingsport, TN (HAAP), the only U.S. facility currently manufacturing RDX and HMX, did not vary greatly, and average yearly concentrations of the nitramines were 2–6 mg/L [121]. Dilution of the effluent by the receiving surface water clearly reduces pollutant concentrations dramatically; reductions to below detection limits are generally found one or two miles downstream [120].

In the manufacture of TNT, numerous nitroaromatic by-products are formed and enter the effluent as plant cleanup and scrubber wastes and as condensates from the evaporative concentration of "red waters" (i.e., selliting

wastes) [122]. Characterization of the steam-volatile trace organics of a representative TNT effluent led to the identification and quantitation of over 30
nitroaromatics. Repeated sampling over a one-year period showed that 2,4- and
2,6-DNT and 1,3-DNB were, in that order, consistently the major effluent
constituents, and together represented approximately 75% by weight of the
total volatile organics [123].

In the manufacture of RDX and HMX the by-products are TAX and SEX.
Levels of TAX are reduced to below detection limits in HAAP effluents by
conventional biological wastewater treatment [121]; RDX levels are reduced to
a lesser extent, and levels of HMX and SEX are not substantially changed by
this treatment.

Soil contamination by munitions compounds has occurred at open-burning
and incinerator sites, and also can result from operational spills and seepage
from landfills and wastewater holding facilities. Pollutant levels can be extremely high in wastewater lagoon beds, the worst case being a lagoon inactive for 20
years, where TNT concentrations were as high as 3000 mg/kg [124]. Soil
surveys of most other munitions plants indicate concentrations many orders of
magnitude lower, even at disposal and open-burning sites. An example of the
latter is sites where RDX concentrations reached 70–80 mg/kg [125].

An important consequence of the burden of munitions compounds in soils is
contamination of ground water. Lysimeter studies have shown TNT migration
to be slow in loamy and sandy soils; the somewhat more rapid movement in clay
soils was attributed to channeling [126, 127]. Nevertheless, a field study reported levels of TNT of > 600 $\mu g/L$ in shallow ground water owing to transloca-
tion from a munitions waste disposal bed [99]. In addition, smaller amounts of
DNT and two microbial reduction products of TNT (see Microbiological
Transformations section) were found. RDX was observed to migrate slowly in
soils [126], but was also present at levels from < 20 to > 700 $\mu g/L$ in ground
water in the vicinity of contaminated sites.

Principal Environmental Fate Processes

In general, two processes can significantly affect the fate and distribution of
munitions pollutants in the environment: microbiological transformations and
photochemical transformations. They will be discussed in the two following
sections. Physical transport from aqueous systems is believed to be unimportant
because both volatilities and sediment adsorption coefficients of the compounds
are low (see Table 4). There is as yet no evidence for other important chemical
transformation processes, such as hydrolysis or oxidation, under environmental
conditions, with the possible exception of a very slow hydrolysis of tetryl
(extrapolated $t_{1/2} \simeq 302 \pm 76$ days at 20°, pH 6.8) [128].

Microbiological Transformations

Microorganisms capable of metabolizing TNT have been reported in soils, composts, muds, and fresh waters and their sediments, as well as in biological waste treatment systems [127, 129–133]. Such transformations have been associated with species from numerous common aerobic and anaerobic bacterial genera including Pseudomonas, Escherichia, Bacillus, Citrobacter, Enterobacter, Klebsiella, Veillonella, and Clostridium. Fungal species representing 98 genera can also transform the compound [134].

While one case of a microorganism growing on TNT as a sole carbon source has been reported [135], the requirement of supplementary nutrients for the metabolism of TNT is now generally well-recognized. Biological systems with added domestic sewage and river sediments have resulted in significant TNT decreases [132, 136], and laboratory cultures with added bacteriological media have resulted in rapid TNT disappearances of 100 mg/L or more [137, 138]. In similar studies, incubation of activated sludge microorganisms with nutrients and ring-labeled ^{14}C TNT resulted in its complete disappearance without formation of labeled CO_2 (mineralization) [139]. Thus, it has been concluded that TNT readily undergoes biotransformation but not biodegradation. The principal products observed from these microbiological transformations of TNT and pathways of their formation are illustrated below. Successive reduction of the nitro groups to amino groups is believed to proceed through hydroxylamine intermediates (not isolated), of which the two shown can also couple to form the observed tetranitroazoxytoluenes [131]. Which products predominate depends on the nature of the microbial preparation, species of microorganisms involved, and conditions under which nitro group reduction takes place. Similar reductive transformations by thermophilic microorganisms under composting conditions were also observed [140]. In a later ^{14}C TNT composting study, however, only traces of reduction products were found; most of the label was associated with non-extractable humic-like materials [141]. Microbial action during treatment of a synthetic munitions-containing wastewater in a semi-continuous activated sludge system resulted in complete disappearance of TNT. Metabolites were sought at various stages of the treatment, and only at a very early stage was 4-amino-2,6-DNT found. It disappeared rapidly, and no other relevant aromatics were detected, but the study had not been designed to determine whether or not biodegradation had taken place [142].

Microorganisms also have the potential to metabolize many of the other nitroaromatic compounds found in TNT production wastewaters. DNT and DNB, unlike TNT, have been shown to be biodegradable. In each case a mixed culture was developed that could grow on and extensively degrade its respective substrate to CO_2 [76, 143]. In similar experiments the more highly nitrogen-substituted wastewater constituents TNB and 3,5-dinitroaniline were not biodegradable [144]. Rather, the slow decreases in their concentrations occurring in the presence of river sediments were attributed to transformations not

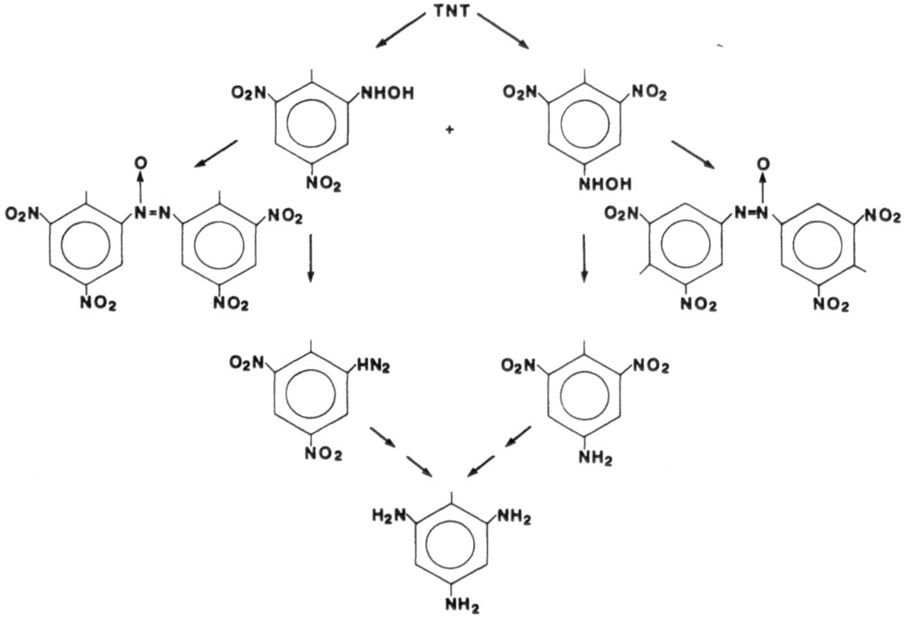

Fig. 2. Microbiological Transformations of TNT

involving ring cleavage; in both cases amino compounds were formed. Biotransformation involving nitro group reduction has also been observed in the case of DNT, which under aerobic conditions gave rise to mixtures of amino- and azoxyaromatics analogous to the TNT reduction products shown in Fig. 2 [145, 146]. Under anaerobic conditions with an exogenous carbon source, the two unstable mononitroso isomers could also be isolated [147]. Little is known about rates of biotransformation of nitroaromatics in the environment except that DNT and TNB, like TNT [124], have been found to persist in soils and wastewater lagoon sediments for many years after a munitions facility became inactive [148, 149].

Despite intensive effort by several groups of investigators, it must be concluded that aerobic biotransformation has little if any effect on the presence of RDX or HMX (or their congeners TAX and SEX) in the environment [76, 150, 151]. On the other hand, certain anaerobic sludge treatment systems with high supplemental nutrient levels were found to effect reductive transformations with relative rates RDX > TAX > HMX > SEX [152–154]. Methanol, formaldehyde, hydrazine, 1,1-dimethylhydrazine, and 1,2-dimethylhydrazine were the major products; sequential pathways involving reduction to nitroso- and hydroxylaminonitramines followed by cleavage reactions were proposed to account for the transformations [152]. Thus, aerobic microorganisms in surface waters appear to be of no consequence in removing nitramines from the aquatic environment, and the significance of anaerobes in removing them from soils and standing water zones or bottom sediments of high nutrient content cannot be more than speculative.

Similarly, NQ is not biotransformed aerobically. However, on prolonged anaerobic treatment of NQ with activated sewage sludge containing high levels of nutrient broth, partial or complete conversion to nitrosoguanidine (NSQ) took place. Small amounts of cyanamide, cyanoguanidine, melamine, and guanidine, which resulted from non-biological degradation reactions of NSQ, were also detected [155]. NQ was not biodegraded in soils except in the presence of high levels of supplemental glucose [156].

In contrast, NG and a number of other aliphatic nitrate esters commonly used in propellant formulations undergo aerobic biodegradation quite readily. Breakdown of NG was found to take place in stages via the isomeric di- and mononitrates, with each successive step proceeding at a slower rate [157]. PGDN, DEGDN, and TEGDN were each found to biodegrade through similar successive steps [158, 159].

Chemical Transformations

Photochemical transformations, due to the effect of sunlight on munitions compounds in environmental waters, are more important than microbiological transformations because their rates are often significantly higher. Indeed, RDX, HMX, and NQ, which strongly resist biodegradation, are readily degraded photochemically. The photochemistry of munitions compounds (notably TNT, and to a lesser extent other nitroaromatics, RDX/HMX, and NQ) has been studied extensively in the laboratory, and because the results, with a few exceptions, have not been published in the open literature, we shall cover them in some detail.

The photolysis of TNT in dilute aqueous solutions, with a medium pressure mercury lamp and a pyrex filter to eliminate wavelengths below 280 nm, gave rise to complex mixtures, owing to further reaction of the primary photoproducts formed initially [160–162] (see Figures 3 and 4). The major primary photoproduct, 2,4,6-trinitrobenzaldehyde (1a), was converted, presumably through the intermediate 2-nitroso-4,6-dinitro-benzoic acid, to the azoxydicarboxylic acid 2a, commonly referred to as the "white compound"; the latter is found as a byproduct in the production of TNT. Smaller amounts of azodicarboxylic acid 3 ("deoxy white compound"), decarboxylated azoxy compound 4, and amide 5 were also formed. In addition, the four isomeric tetranitroazoxytoluenes 6a–d were present as minor products [160].

Two other important photoproducts were 2-amino-4,6-dinitrobenozic acid (7a) and 4,6-dinitro-1,2-benzisoxazole (4,6-dinitroanthranil, 8a). The latter was photolabile and was converted on further irradiation to the tetracyclic compound 9. A minor photoproduct, 2,4,6-trinitrobenzyl alcohol, was also photolabile and gave rise to mixtures of aldehyde 1a, azoxydialdehyde 2b, and 3.

While 1a was also the principal sunlight-promoted TNT photoproduct in

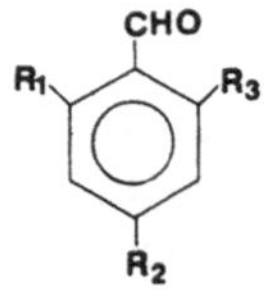

1a $R_1=R_2=R_3=NO_2$

1b $R_1=NH_2$, $R_2=NO_2$, $R_3=H$

1c $R_1=R_2=NO_2$, $R_3=H$

1d $R_1=R_3=NO_2$, $R_2=H$

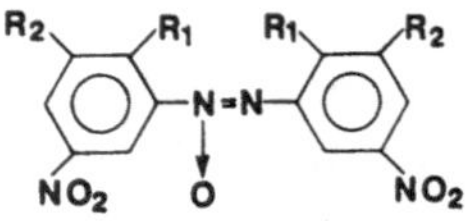

2a $R_1=CO_2H$, $R_2=NO_2$

2b $R_1=CHO$, $R_2=NO_2$

2c $R_1=CO_2H$, $R_2=H$

2d $R_1=CHO$, $R_2=H$

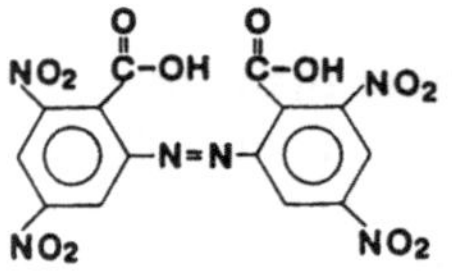

3

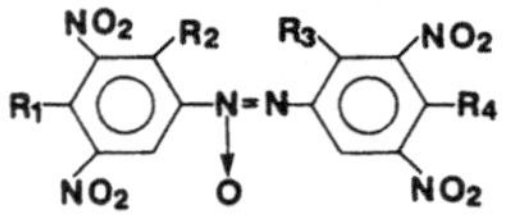

6a $R_1=R_4=CH_3$, $R_2=R_3=H$

6b $R_1=R_4=H$, $R_2=R_3=CH_3$

6c $R_1=R_3=H$, $R_2=R_4=CH_3$

6d $R_1=R_3=CH_3$, $R_2=R_4=H$

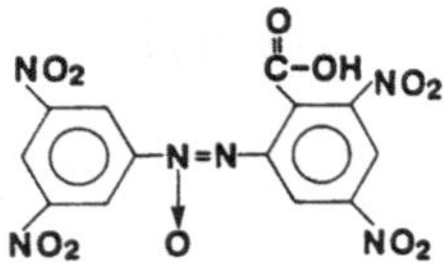

4

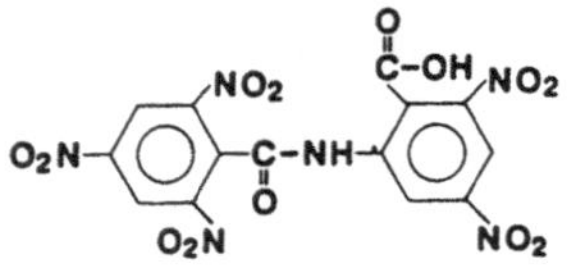

5

Fig. 3. Structures of Compounds 1–6

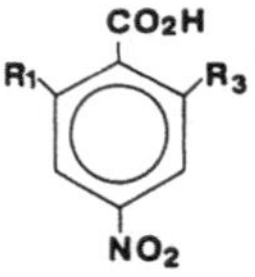

7a $R_1=NH_2$ $R_2=H$

7b $R_1=NH_2$ $R_2=H$

7c $R_1=NO_2$ $R_2=H$

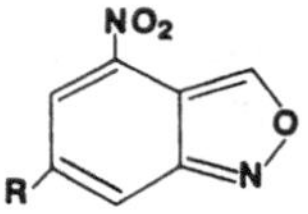

8a $R=NO_2$

8b $R=H$

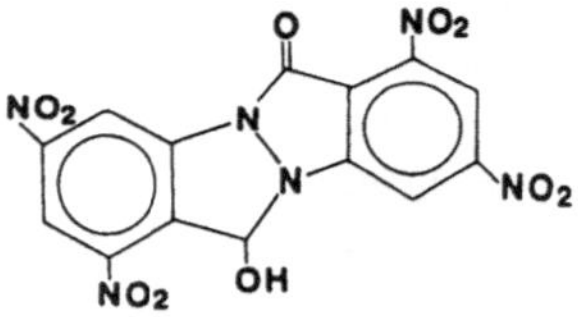

9

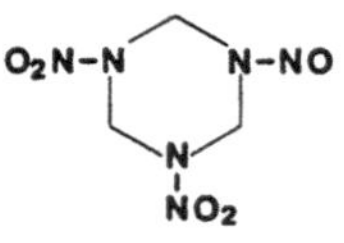

18

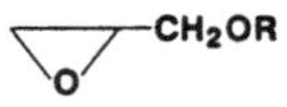

19a $R=NO_2$

19b $R=H$

Fig. 4. Structures of Compounds 7–9, 18, 19

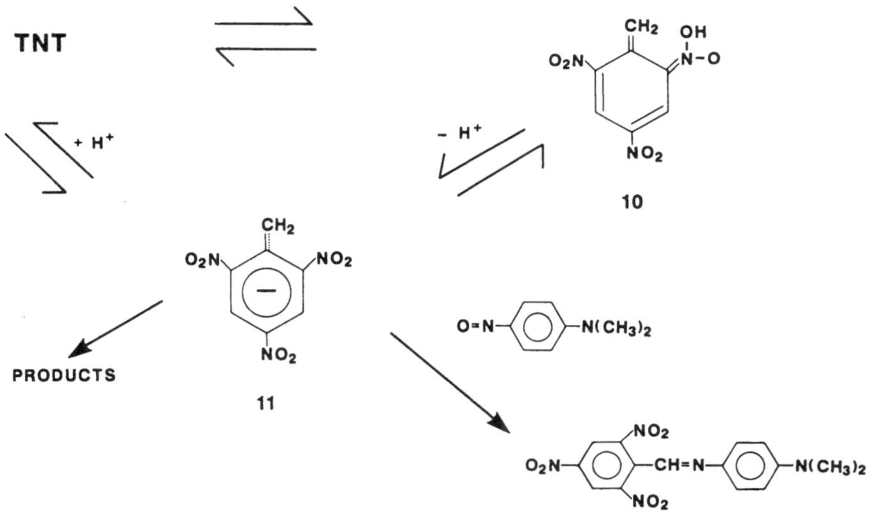

Fig. 5. Mechanism of Photolysis of TNT

distilled water, TNB (a minor product in the laboratory studies) was the major photoproduct observed in a similar study of the effect of sunlight on TNT in natural river water [163]. This photostable compound then underwent slow biotransformation via nitro group reduction to 3,5-dinitroaniline.

Similar laboratory studies on other nitroaromatics have shown that photolysis of 2,4-DNT follows a similar course: the major products isolated were 2-amino-4-nitrobenzoic acid (7b), 2-amino-4-nitrobenzaldehyde (1b), and azoxydicarboxylic acid 2c; smaller amounts of 2,4-dinitrobenzaldehyde (1c), 2,4-dinitrobenzoic acid (7c, DNBA), and azoxydialdehyde 2d were present as well [161]. Photolysis of 2,6-DNT gave rise to unstable mixtures; little is known of the identities of the products except that 2,6-dinitrobenzaldehyde (1d) and 6-nitroanthranil (8b) were probably present, in addition to certain azoxy- and azobenzenes [161]. Evidence for the presence of oligomers of azoxy/azo-benzenes (3 or more aromatic rings) was also found.

Photooxidation of 2,4-DNT in the presence of large excesses of hydrogen peroxide resulted in oxidation to the corresponding acid (via the alcohol and aldehyde), decarboxylation, and hydroxylation of the resulting DNB. Subsequent ring cleavage to mixtures of low molecular weight acids and aldehydes was observed [164].

As to the mechanism of phototransformation of TNT, the rate of disappearance of TNT was inversely proportional to pH, and in D_2O the deuterium content (found only in the methyl group) of the recovered TNT was substantially greater at the lower pHs [162, 165]. Thus the reaction is believed to proceed by intramolecular hydrogen transfer in the triplet excited state to the aci-quinoid species **10**, followed by deprotonation to the intermediate TNT anion **11**, which can react further or reprotonate. Further evidence for the intermediacy of **11** was a trapping experiment with p-nitroso-N,N-dimethylaniline to form a Schiff base [165].

Rate enhancements of 10 to 100-fold for TNT photolysis in natural waters have been observed and may be attributed to the action of humic acids as triplet sensitizers [166]. A study of 19 nitroaromatics under similar conditions showed enhancements of 2- to 26-fold, with the greatest in cases of nitro groups with *ortho* methyls. Thus, 2,6-DNT showed a greater enhancement than 2,4-DNT [167]. A recent investigation to probe the nature of excited species in humic-sensitized photoreactions of aquatic pollutants showed that the key steps in the reactions studied (photoisomerization of 1,3-pentadiene and photooxidation of 2,5-dimethylfuran) involved the transfer of electronic energy from triplet states of the humic substances, some sufficiently high to transfer energy to nitro-aromatics [168].

Laboratory studies have shown that UV photolysis of nitramine munitions in aqueous solutions with an unfiltered medium-pressure mercury lamp is exceedingly facile (completion within 7–10 min), and the ultimate products are nitrate, nitrite, ammonia, and formaldehyde [169, 170]. Carbon dioxide was formed only if the mixtures were treated with ozone during or after photolysis [170]. Photolysis alone has been viewed as a possible cost-effective small-scale alternative to activated carbon for removal of nitramine munitions from waste-water effluents [171]. Studies of the early stages of photolysis of RDX have led to the isolation and characterization of some of the intermediate organic nitro compounds, and have contributed substantially to our understanding of these complex photoprocesses [170, 172, 173]. The results, shown below, indicated that several different pathways may be operative. Photolysis in ethanol revealed one probable course of the reaction in water: in ethanol the products isolated were ethyl nitrate (**12a**), diethyl formal (**13a**), and methylenedinitramine (**14**) [172]. In water a similar pathway would lead to nitrate and formaldehyde (**13b**); **14** decomposes in water to formaldehyde and nitrous oxide, which have also been observed [173]. Three other products isolated from aqueous photolysis were assigned structures **15**, **16** and **17** on the basis of their mass spectra, and were viewed as arising by a different pathway, possibly via a transient ni-trodihydrotriazine [170]. Evidence for formation of the nitroso compound **18** was also found [173].

Aqueous solutions of both RDX and HMX were photolyzed slowly by sunlight to mixtures of the same ultimate products found in the UV laboratory study of RDX; rates observed were in terms of days rather than minutes, and RDX reacted 2–3 times faster than HMX [150].

Tetryl in aqueous solution was also photolyzed slowly by sunlight (20 days to essential completion); N-methylpicramide, nitrate, and nitrite were the major products [128]. A slower hydrolysis (90 days to essential completion) was observed at pH 9 in buffered solutions protected from light; the major products were methylnitramine and picrate ion [128].

NQ in aqueous solution was photolyzed even more readily by UV irradiation (254 nm) than RDX, and the rates of photodegradation by sunlight in natural waters were comparable to RDX [74, 174]. Laboratory studies of the UV photolysis of NQ, summarized in Table 5, showed the products to be primarily guanidine, urea, and nitrite, with lesser quantities of cyanoguanidine, ammonia, and nitrate. Nitrosoguanidine (NSQ) was found to be a transient

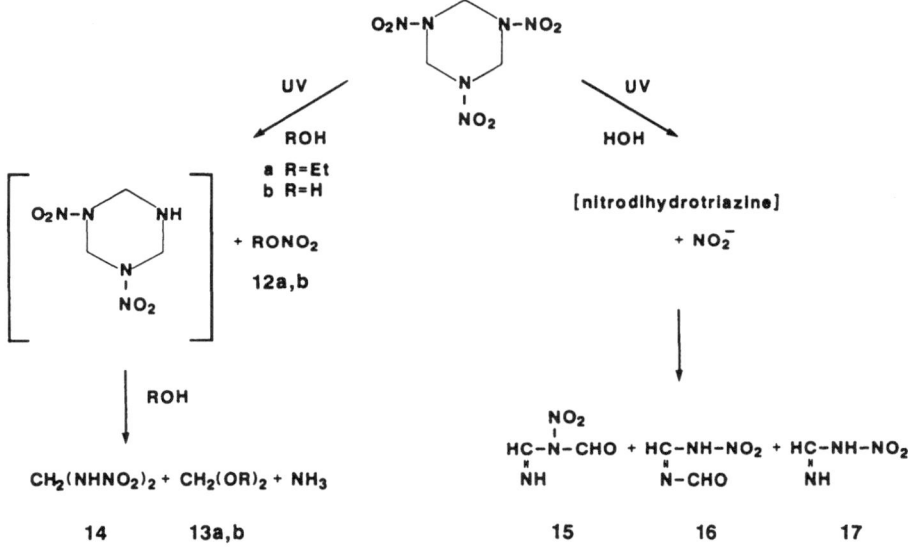

Fig. 6. Photolysis Pathways of RDX

Table 5. Photolysis products of nitroguanidine and nitrosoguanidine[a]

Product	Reactant/Conditions			
	NQ		NSQ	
	unbuffered	pH 10	unbuffered	pH 10
NSQ				
maximum	0.02	0.008	1.00	1.00
terminal	< 0.002	< 0.003	ND[b]	ND
Guanidine	0.37	0.05	0.90	0.15
Cyanoguanidine	0.05	0.015	0.03	0.01
Urea	0.29	0.19	NM[c]	0.22
Nitrite N	1.10	0.65	1.4	0.9
Nitrate N	0.17	ND	ND	ND
Ammonia N	0.10	NM	0.06	NM
Nitrogen gas[d]	ND	0.23	NM	NM

[a] Quantities are expressed as mole product/mole reactant; except where noted, initial reactant concentrations were $0.5 - 1 \times 10^{-3}$ M.
[b] Not detected.
[c] Not measured.
[d] Initial reactant concentrations were 10^{-2} M.

intermediate, which was itself photolyzed largely to guanidinium nitrite. At high pH distribution of photolysis products was different, and nitrogen gas was formed [174].

While nitrate esters in aqueous solution are not significantly photolabile, they are readily hydrolyzed at higher pHs. Because desensitization of wastewater from NG manufacture by chemical treatment with calcium hydroxide has been viewed as a viable treatment option, the kinetics and the products formed on hydrolysis of NG and each of the isomeric DNGs and MNGs were extensively studied [175–177]. The results show that chemical hydrolysis does not follow the stepwise, successively slower pathway found for microbiological systems. Rather, NG was hydrolyzed more slowly than the DNGs or MNGs to a complex mixture of products, notably nitrate, nitrite, and oxalate [177]. 1,3-DNG was hydrolyzed to glycidyl nitrate (**19a**), and 1,2-DNG was shown to isomerize to 1,3-DNG before hydrolysis to the same product [176]. Similarly, 2-MNG was isomerized to 1-MNG before hydrolysis; the only product isolated was nitrate. Glycidol (**19b**) was isolated only on brief treatment of 2-MNG with 30% sodium hydroxide [175]. In every case the calcium hydroxide hydrolyses were kinetically second order (or pseudo-first order with excess hydroxide). The chemically generated epoxides **19a, b** were found to be biodegradable [178].

Metabolism

In mammalian systems the principal metabolites of TNT are 2-amino-4,6-DNT (2ADNT) and 4-amino-2,6-DNT (4ADNT); smaller amounts of 2,4-diamino-6-nitrotoluene and trace amounts of 2,6-diamino-4-nitrotoluene are also formed. This pattern has been found repeatedly in urine of rats, blood of rabbits, and urine of munitions workers [179, 180]. In one reported plant metabolism study 4ADNT and, to a lesser extent, 2ADNT were found throughout the plant [181]. We have estimated first order rate constants for clearance of ^{14}C-TNT in 4 mammalian species based on data for tissue distribution and excretion of the radiolabel [182]; the half-lives therefrom were 4–7 hours.

Clinical investigations of NG have shown that the major metabolites are 1,2- and 1,3-DNG; recent procedures for their determination in picogram amounts from blood plasma have been published [183–185]. Both TNT and NG are readily absorbed through the skin and analysis of metabolites as well as the parent compounds may be a promising forensic tool [179, 186].

Studies of 2,4- and 2,6-DNT in rats [187, 188] and in humans [189, 190] show that reduction of the nitro groups is not the predominant metabolic pathway for either isomer; rather, oxidation of the methyl group is favored. For 2,4-DNT in both species, 2,4-dinitrobenzoic acid (DNBA) is the major metabolite. In rats, 2,4-dinitrobenzyl alcohol (DNBalc, excreted as glucuronide) predominated over 2-amino-4-nitrobenzoic acid (**7b**) [188], while in humans the reverse was observed. For 2,6-DNT in rats, a roughly 1:1:1 mixture of the three corresponding metabolites resulted [187], whereas in humans roughly equal amounts of DNBA and DNBalc were found and 2-amino-6-nitrobenzoic acid was not detected.

In rats ^{14}C-RDX was metabolized in the liver; after 4 days 40–50% of the label was recovered as CO_2 and up to 35% was found in the urine, largely as unidentified metabolites [191, 192]. Only 10% of the label remained in the carcass. Estimates of clearance rates from these data resulted in half-lives of 24–30 hours. In one recent case of human ingestion [193] the question of metabolism was not addressed, but quantitative HPLC determinations showed maximum RDX concentrations of 11 mg/mL in the serum after 24 hours, and 38 mg/L in the urine after 48 hours [194].

In rats ^{14}C-NQ was not metabolized but was quantitatively recovered unchanged in the urine [195]. The exceedingly low acute toxicity of the compound (Table 6) is in accord with this observation.

In humans and animals the major metabolite of PA has long been known to be picramic acid (2-amino-4,6-dinitrophenol, PrA). The metabolism of each of these compounds has recently been studied in rainbow trout [79]. In separate experiments, PA was reduced to PrA and PrA was oxidized to PA, and in both cases mixtures of the two glucuronides were obtained. Bioconcentration factors (BCF) of < 1 for both PA and PrA in trout epaxial tissue were determined in this study, and may be attributed to facile conjugation and excretion.

Mammalian Toxicology (Including Human)

The effects of nitrate esters (notably NG, EGDN, PGDN, and PETN) on human and animal health and physiology have long been known and are detailed in a recent comprehensive review [196]. In humans, the most prominent manifestations of NG and EGDN toxicity are severe headaches and adverse cardiovascular effects. PGDN at levels twice that of the recommended threshold limit value (TLV) (see p. 223 and Table 11) causes headache, but no cardiovascular or neurological effects were found in studies of occupationally exposed torpedo maintenance workers [196–198]. Data are sparse for PETN, but it appears to be relatively non-toxic [196].

Toxic manifestations of TNT in humans, notably aplastic anemia and toxic hepatitis, have been known for at least 70 years, and similar conditions have been observed in animals [199, 200]. Other related nitroaromatics (DNT, DNB, PA, and tetryl) produce similar effects at high doses; PA and tetryl also cause contact dermatitis [199]. Little information is available for TNB.

RDX has adverse effects on the central nervous system (CNS) in mammals [201, 202], and produces convulsions and/or unconsciousness on exposure by inhalation or by ingestion [193, 203–205]. In cases of human intoxication, recoveries have been complete within days or, at worst, several months after exposure. Medical evaluation of workers chronically exposed to RDX at low levels (not greater than 1.57 mg/m^3) has shown no abnormalities of hematologic, hepatic, or renal systems [206]. Similar CNS effects have been demonstrated for HMX in rodents, but at significantly higher doses [207]. NQ has not been found significantly toxic to rodents [208–210].

Acute toxicities to rodents are summarized in Table 6, and Table 7 summarizes the no-observed-effect-levels (NOEL) or no-observed-adverse-effect-levels (NOAEL) estimated from certain of the longer term chronic and subchronic studies.

Evidence for carcinogenicity of TNT and RDX from mammalian studies is limited; consequently both compounds have been classified by the U.S. Environmental Protection Agency (EPA) as class C carcinogens (see Table 11). DNT, on the other hand, is well recognized as carcinogenic to rodents. Because much of the mammalian toxicity testing of DNT has been carried out on mixtures of the 2,4- and 2,6-isomers, there has been considerable confusion as to the roles of the individual isomers in carcinogenesis. Recent work, however, has established that in rats 2,6-DNT appears to be a complete hepatocarcinogen (initiator and promotor) while 2,4-DNT is an apparent pure promotor and is considered a much less potent carcinogen [211].

As to the question of carcinogenicity of DNT in man, the results of a recent retrospective cohort mortality study involving 457 former personnel occupationally exposed to DNT at Joliet and Radford Army ammunition plants during

Table 6. Acute toxicities of munitions compounds to rodents

Compound	Rat mg/kg[a] (route)	Mouse mg/kg[a] (route)	Reference
TNT	800–1300 (oral)	600–1000 (oral)	[214]
DNT	200–800 (oral)	1200–2000 (oral)	[215]
NG	500–900 (oral)	500–1200 (oral)	[216]
	100–110 (ip)	100–200 (ip)	[216]
	25–32 (iv)	10–18 (iv)	[216]
	500–600 (sc)	30–500 (sc)	[216]
RDX	40–300 (oral)	60–500 (oral)	[201]
		19 (iv)	[201]
HMX	6250 (oral)	2300 (oral)	[217]
		634 (sc)[b]	[207]
NQ	> 5000 (oral)	5000 (oral)	[195,208,209]
TNB	450 (oral)	572 (oral)	[63]
		32 (iv)	[63]
DNB	83 (oral)	200 (ip)[c]	[63]
DEGDN	700–1000 (oral)[d]	1300–1400 (oral)[d]	[218]
	777 (oral)		[63]
EGDN	616 (oral)		[63]
PGDN	250 (oral)		[63]
	479 (ip)	1047 (ip)	[63]
	463 (sc)	1208 (sc)	[63]
PA		100 (oral)[e]	[63]
Tetryl		5000 (sc)[f]	[63]

[a] LD50 unless noted otherwise.
[b] In rabbit.
[c] LDLo.
[d] LD100.
[e] LDLo in guinea pig.
[f] LDLo in dog.

Table 7. NOEL or NOAEL for selected munitions compounds estimated from chronic and subchronic toxicity data

Compound	Duration of Test	NOEL/NOAEL (mg/kg/day)	Species	Reference
TNT	13 wk	5.0[a]	rat	[219]
	13 wk	1.4–1.45[b]	rat	[220]
	13 wk	1.45–1.6[b,c]	mouse	[220]
	13 wk	0.2	dog	[220]
	26 wk	0.5[d]	dog	[221]
	2 yr	0.4	rat	[222]
DNT	1 yr	13.5[e]	mouse	[223, 224]
	1–2 yr	0.6[b]	rat	[223, 225]
	2 yr	0.2	dog	[223]
NG	13 wk	25.5	rat	[226]
	1 yr	1.0	dog	[227]
	2 yr	3–4[b]	rat	[227]
	2 yr	10–11[b,f]	mouse	[227]
RDX	13 wk	15	rat	[202]
	13 wk	80	mouse	[202]
	13 wk	1.0	monkey	[202]
	2 yr	0.3	rat	[228]
	2 yr	1.5	mouse	[229]
HMX	13 wk	50–115	rat	[207]
PA	2 yr	25[g]	rat	[230]
NQ	13 wk	316	rat	[210]
DNB	16 wk	0.75	rat	[231, 232]
PETN	1 yr	2	rat	[196]

[a] Changes in liver weight.
[b] Males and females were given slightly different doses.
[c] Enlarged spleens and hearts.
[d] Mild liver lesions observed in 7 of 12.
[e] LOAEL.
[f] Pigmented intracellular granules in the liver.
[g] Calculated from concentration in feed (500 ppm), assuming standard animal weight (200 g) and feed consumption (10 g/day)[63].

the 1940s and 1950s provide some significant information [212]. Workers at Joliet were exposed to crude DNT (containing 19% 2,6-isomer) while at Radford the DNT contained no more than 1% 2,6-isomer. No deaths from any type of cancer in excess of the "standard mortality ratio" or percentage of expected deaths were found in either of the two groups. On the other hand, elevated mortality due largely to heart and circulatory diseases of atherosclerotic origin was found in both groups. The data suggested a correlation of increased mortality with duration and intensity of exposure.

A similar mortality study involving workers occupationally exposed to TNT and EGDN in the Swedish dynamite industry showed double the expected number of deaths from cardio-cerebrovascular causes [213].

Aquatic Toxicology

TNT, DNT, and the many other nitroaromatic by-products of TNT manufacture have been extensively investigated both singly and in mixtures. Acute studies utilized a variety of fish, invertebrate, and algal species [233–235], and the test conditions (static or flow-through), concentration determinations (nominal or measured), and biological endpoints were not generally uniform. Endpoints were usually determined as 48- and 96-hour LC50s, or as time-independent concentrations termed as "incipient" LC50s and defined as the concentrations above which 50 percent of the test organisms would not survive indefinitely [236]. Table 8 summarizes the results of acute toxicity tests conducted under static conditions for a variety of fish and invertebrates, and Table 9 is a similar summary of tests conducted under flow-through conditions and including incipient as well as time-dependent LC50 determinations.

Comparison of the 96-hr LC50s for the four fish species shows significantly higher values for the flow-through conditions, except in the bluegill case. In general under the flow-through conditions, incipient LC50s were consistently lower than the 96-hr values. For the fish, it is apparent that photolysis reduced toxicity in the static tests. Variation of water quality parameters such as hardness, pH, and temperature had little effect on acute toxicities of TNT [234].

Table 8. Acute toxicity of TNT to selected species of fish and invertebrates under static conditions[a]

Species	Exposure Time	LC50 (mg/L)[b]	
	(hr)	non-photolyzed	photolyzed[c]
Fathead minnow (*Pimephales promelas*)	96	2.9 (2.6–3.2)	12.8 (11.6–13.9)
Channel catfish (*Ictalurus punctatus*)	96	2.4 (2.0–2.7)	5.5 (4.8–6.4)
Rainbow trout (*Salmo gairdneri*)	96	0.8 (0.7–1.0) 1.4 (1.2–1.8)[d]	13.9 (9.0–18.0)
Bluegill (*Lepomis macrochirus*)	96	2.6 (2.3–2.9) 3.4 (3.1–3.7)[d]	18.3 (17.2–19.5)
Water flea (*Daphnia magna*)	48	11.7 (10.9–12.6)	16.5 (14.3–18.6)
Scud (*Hyalella azteca*)	48	6.5 (5.6–7.5)	6.7 (5.0–8.9)
Midge (*Tanytarsus dissimilis*)	48	27.0 (22.0–33.0)	25.2 (17.3–30.6)
Earthworm (*Lumbriculus variegatus*)	48	5.2 (4.5–6.0)	19.2 (16.0–22.3)

[a] Data from Ref. [234]. Concentrations nominal.
[b] Data in parentheses represent 95% confidence limits.
[c] LC50s for photolyzed TNT are based upon concentration of TNT before irradiation with a Pyrex-filtered medium pressure mercury lamp.
[d] Test solutions aerated.

Table 9. Acute toxicity of TNT to selected species of fish and invertebrates under flow-through conditions[a]

Species	LC50 (mg/L) 48-hr	96-hr	Incipient[b]	Duration (hr)
Fathead minnow	5.9	3.7	1.5 (0.9–2.5)	384
Channel catfish	5.6	3.3	1.6 (0.9–3.0)	288
Rainbow trout	2.0	2.0	1.9 (1.3–3.3)	240
Bluegill	2.6	2.5	1.4 (0.8–2.5)	312
Water flea	> 4.4	1.2	0.2 (0.1–1.0)	192
Earthworm	> 29.0	> 29.0	13.9 (12.6–15.1)	336

[a] Data from Ref. [234]. Concentrations measured.
[b] Data in parentheses represent 95% confidence limits.

The various nitroaromatics resulting from the TNT manufacturing process have been tested, both as individual compounds and in mixtures of condensate water, for acute toxicity to fathead minnows and water fleas [233]. The results for selected compounds are summarized in Table 10. It is evident that 2,4- and 2,6-DNT, the most abundant constituents of condensate water, are among the

Table 10. Acute toxicity of selected constituents of condensate water to fathead minnows and water fleas under static conditions[a]

Compound[b]	Fathead minnow 96 hr LC50 (mg/L)[c]	Water flea 48 hr LC50 (mg/L)[c]
2,4-DNT	32.8 (27.3–38.0)	47.5 (29.5–99.7)
2,6-DNT	18.5 (17.2–20.2)	21.8 (19.3–24.6)
1,3-DNB	7.0 (5.8–8.1)	49.6 (42.5–59.2)
	16.8[d]	27.4 (24.0–31.4)[d]
3,5-DNT	22.6 (13.4–27.1)	45.2 (42.4–48.4)
3,4-DNT	1.5 (1.1–1.8)	3.7 (1.0–5.4)
2,3-DNT	1.8 (1.5–2.1)	4.7 (3.0–5.9)
2,5-DNT	1.3 (1.1–1.4)	3.1 (2.2–3.8)
2,4,6-TNT	2.9 (2.6–3.2)	11.7 (10.9–12.6)
2,3,6-TNT	0.1 (0.10–0.13)	0.8 (0.4–1.2)
1,3,5-TNB	1.1 (1.0–1.2)	2.7 (2.4–3.1)
	0.5 (0.4–0.6)[d]	3.0 (2.6–3.4)[d]
3,5-Dinitroaniline	21.8 (19.1–31.3)	15.4 (13.5–18.0)
	21.2 (15.1–29.9)[d]	13.8 (12.9–14.8)[d]
2-Amino-3,6-DNT	0.9 (0.7–1.2)	2.5 (0.7–6.5)
3-Amino-2,4-DNT	12.2 (10.8–13.4)	9.6 (5.5–11.0)
Condensate water	185 (164–207)	

[a] Data from Ref. [233] unless noted otherwise. Concentrations nominal.
[b] Listed in order of decreasing occurrence in condensate water; the first 3 constitute 43%, 22%, and 12%, respectively, of the total organic content.
[c] Data in parentheses represent 95% confidence limits.
[d] Data from Ref. [235].

least toxic. The table further shows a remarkable correlation of substitution pattern with toxicity among the DNT isomers. Isomers with *ortho* (2,3 and 3,4) and *para* (2,5) nitro groups are about 10–20 times more toxic than those with *meta* (2,4, 2,6, and 3,5) nitro groups. This effect is also manifest to an even greater degree in the comparative toxicities of 2,3,6- and 2,4,6-TNT. The somewhat greater toxicities observed for condensate water relative to its major constituent are readily attributable to the presence of these highly toxic minor components.

Full life-cycle chronic tests have been made with fathead minnows and water fleas for TNT, DNT, and condensate water [237, 238]. Early life stage tests (30, 60, and 90 days) were made with fathead minnows, channel catfish, and rainbow trout. Reproductive parameters included spawning, egg hatching, and fry survival and growth. Results of the chronic studies showed adverse effects even at the lowest concentrations tested (0.04 mg/L for TNT and 0.28 mg/L for DNT). In three-generation tests with fathead minnows, adverse effects were generally more pronounced in the later generations. In the early life stage tests substantially higher concentrations were required for observation of adverse effects.

Most of the available information on toxicity of TNT (and in a few cases, DNT) to aquatic plants deals with algal species exposed from 16 hours [239] to 15–17 days [240]. Concentrations at least as high as the incipient LC50s observed for fish and water fleas (Table 9) were required for adverse effects. One study reported a value of 1.0 mg/L for growth inhibition of the vascular species *Lemna perpusilla* (duckweed), and death of the plant at 5 mg/L [241]. Only a single study of the effect of TNT on the growth of a terrestial plant, yellow nutsedge (*Cyperus esculentus* L.), in hydroponic culture has been published [181]. Adverse effects were observed at 5 mg/L, the lowest concentration studied.

Information regarding adverse effects of RDX to aquatic organisms is sparse. One study cited 96-hr LC50s of 4–6 mg/L (static) and 7–13 mg/L (flow-through), and incipient LC50s of 5–11 mg/L (flow-through) in the four fish species listed above (Tables 8, 9) [242]. The validity of these and other data reported in the study has been questioned [201, 243]. Acute toxicities of mixtures of RDX and TNT (1.6:1) were not significantly different from those observed for TNT alone [234]. No toxic effects were observed for HMX up to its solubility limit (5 mg/L).

NQ is essentially non-toxic to aquatic organisms [244]. Acute tests with four fish species and five invertebrates showed no LC50s up to the solubility limits (ranging from 1.7 to 3.0 g/L at temperatures between 12 and 22 °C). Chronic toxicity was also exceedingly low. A 42-day early life stage test with rainbow trout (the species found most sensitive) showed significant differences from the controls only at saturation. The mixture produced by photolysis of NQ, however, was about 100 times more toxic to fathead minnows and water fleas (LC50 and EC50 about 30 mg/L in each case). Of the major NQ photolysis products (Table 5), guanidine and/or nitrite may account for the observed toxicities.

NG, on the other hand, is quite toxic to fish, with 96-hr acute (both static and flow-through) values of 1.7–4 mg/L for four species. Chronic studies in the fathead minnow showed adverse effects from as little as 0.2 mg/L. Four species of invertebrates were less sensitive, with 48 hour EC50s in the range 20–55 mg/L [77].

An acute toxicity study of DEGDN to four species of fish and four species of invertebrates showed it to be relatively non-toxic [245]. The most sensitive invertebrate (water flea) was substantially more sensitive (48-hr EC50 90 mg/L) than the most sensitive fish (96-hr static LC50 258 mg/L, bluegill).

Acute toxicities (96-hr LC50) of 110 and 46 mg/L for PA and picramide, respectively, to rainbow trout have been reported [246]. A 42-day chronic study in rainbow trout at concentrations < 0.001 LC50 of PA and picramide showed no growth inhibition, but in both cases > 80% of the fish developed petechial lesions [247]. Similar 42-day studies with PA and picramic acid concentrations as low as 0.05 and 0.02 mg/L, respectively, showed significant uptake and inhibition of growth in oysters (*Crassostrea virginica*) [247, 248].

While insufficient toxicological data are available at present to meet the most recent EPA guidelines [249] for numerical estimates of water quality criteria to assure protection of the aquatic biota, some interim estimates have been made on the basis of the available data for selected munitions compounds. The criteria specify two concentrations, a criterion maximum concentration (CMC) and a criterion continuous concentration (CCC), which should not be exceeded. Estimated values for TNT were, respectively, 0.56 and 0.04 mg/L [214]; for DNT, 5.5 and 0.2 mg/L [215]; and for NG 0.86 mg/L (CMC only) [216].

Water Quality Criteria and Standards for Human Exposure

The ultimate products of mammalian toxicological evaluation and human epidemiology of environmental pollutants are official opinions, criteria, or standards for human exposure. Such regulatory or quasi-regulatory values have been developed in the U.S. for ten of the compounds discussed above (Table 11).

Threshold limit values (TLV) have been issued by the American Conference of Industrial Hygienists as recommended occupational health standards; TLVs have generally been accepted as permissible exposure limits (PEL) by the U.S. Occupational Safety and Health Administration (OSHA), though often after lag times of several years.

The EPA Office of Drinking Water (ODW) publishes health advisories (HAs) that are not binding according to federal law; nevertheless individual states have included values from HAs in their regulations as if such numbers had been fully validated. The drinking water equivalent levels (DWEL) cited by ODW in HAs (Table 11) are based on the assumptions of non-carcinogenicity of the compounds and of drinking water as the sole source of human exposure.

Table 11. Proposed and official regulatory values and classifications

Compound	ACGIH TLV[a] TWA[c] ppm[g]	mg/m³	OSHA/NIOSH PEL/REL[b] TWA[c] ppm[g]	mg/m³	STEL[d] mg/m³	HA DWEL[e] mg/L	Potency factor[f] (mg/kg/d)⁻¹	Carcinogen class[h]
TNT	—	0.5	—	1.5, (0.5)[i]	—	0.020	0.031	C
2,4-DNT	—	1.5	1.5	—	—	—	0.311[j]	B2
DNB	0.15	1	0.15	1	—	—	—	—
RDX	—	1.5	—	—	—	0.10	0.11	C
HMX	—	—	—	—	—	1.8	—	D
Tetryl	—	1.5	—	—	—	—	—	—
PA	—	0.1	—	0.1	0.3	—	—	—
NG	0.05	0.5	0.2	2	(0.1)[k]	0.005	0.0166	C[l]
EGDN	0.05	0.3	0.2	1	(0.1)[k]	—	—	—
PGDN	0.05	0.3	—	—	—	—	—	—

[a] American Conference of Governmental Industrial Hygienists threshold limit value [251].
[b] Occupational Safety and Health Administration (OSHA) permissible exposure limit and National Institute of Occupational Safety and Health (NIOSH) recommended exposure limit [252].
[c] Time weighted average for an 8-hr/day, 5-day workweek [251].
[d] Short-term exposure limit, usually 15 min, not to be repeated more than 4 times per day or at intervals of < 60 min [251].
[e] EPA Office of Drinking Water (ODW) drinking water equivalent level for lifetime exposure. DWELs have been published in the respective Health Advisories (HAs) [202, 207, 226, 250].
[f] Carcinogenic potency factor, q_1^*, calculated by ODW. Multiplying q_1^* in (mg/kg/day)⁻¹ by the daily exposure level in mg/kg/day gives the lifetime cancer risk, R. Thus if one assumes an acceptable R of 10^{-6} (one excess cancer from a lifetime exposure of one million people), an acceptable daily exposure level is $10^{-6}/q_1^*$ mg/kg. Unless otherwise noted, values appear in the respective HAs [202, 226, 250].
[g] Parts per million in the vapor phase, i.e., molecules per million molecules of air. At 25 °C, ppm = (mg/m³ × 24)/MW.
[h] Group classifications as follows have been made by USEPA.
B2: Probable human carcinogen; usually on the basis of adequate evidence in animals and inadequate evidence or insufficient data in humans.
C: Possible human carcinogen; limited evidence of carcinogenicity in animals and insufficient data in humans.
D: Not classified as carcinogen; no evidence from animal studies.
[i] Currently proposed by OSHA.
[j] ODW has not yet issued HAs for 2,4- or 2,6-DNT. However, a tentative value of 0.311 (mg/kg/day)⁻¹ for the carcinogenic potency factor q_1^* for 2,4-DNT has been cited [253]. Assuming a value of 10^{-6} for R (see footnote f), the acceptable exposure limit is $10^{-6}/0.311$ or 3.2×10^{-6} mg/kg/day. Thus, for a 70 kg individual consuming 2 L of water per day, a concentration of 0.00011 mg/L is calculated.
[k] Currently recommended by NIOSH for 20-min STEL.
[l] Authors' estimate; the EPA has not yet classified NG.

Additional safety factors are applied in the calculation of lifetime advisory limits to allow for other routes of exposure and for the uncertainty as to the no-effect levels of compounds designated by EPA as class C carcinogens (see Table 11) [202, 226, 250].

It must be emphasized, in conclusion, that both our understanding of the toxicology of environmental pollutants and the application or conversion of toxicological data to criteria for decision-making are still in the early stages of development.

Abbreviations/Symbols

Symbol	Definition
BCF	Bioconcentration factor
ΔF_f	Free energy change in going from solid to liquid state
ΔH_m	Heat of fusion
K	Degrees Kelvin
K_H	Henry's law constant
K_d	Actual partition coefficient of a chemical between soil and water
K_{oc}	Soil organic carbon/water partition coefficient
f_{oc}	Fraction of organic carbon in soil; $f_{oc} = K_d/K_{oc}$
K_{ow} (or P_{ow})	Octanol-water partition coefficient
MW	Molecular weight
P	Vapor pressure
$P_{(l)}$	Vapor pressure in liquid state
$P_{(s)}$	Vapor pressure in solid state
R	Gas constant
S	Solubility in water
ΔS_f	Entropy of fusion
T	Temperature in degrees Kelvin
T_m	Melting point in degrees Kelvin unless specified otherwise

References

1. US Dept of the Interior, Bureau of Mines (1987) mineral industry surveys, Apparent consumption of industrial explosives and blasting agents in the United States, 1986, Washington, DC
2. *Lindner V* (1980) Explosives and propellants, in: Grayson M, Eckroth D (eds) Kirk-Othmer encyclopedia of chemical technology, 3rd ed, vol 9, Wiley, New York, p 561
3. *Kaye S M* (ed) (1980) [TNT], purification, in: Encyclopedia of explosives and related items, vol 9, Dover, NJ, Large Caliber Weapon Systems Lab, US Army Armament Research and Development Command. AD-A097595, p T243
4. *Kaye S M* (ed) (1980) [Toluene and derivatives], the dinitrotoluene isomers (DNT), in: Encyclopedia of explosives and related items, vol 9, Dover, NJ, Large Caliber Weapon Systems Lab, US Army Armament Res Develop Com AD-A097595, p T307
5. le*Spanggord R J, Mill T, Chou T W, Mabey W R, Smith J H, Lee S* (1980) Environmental fate studies on certain munition wastewater constituents, Final report, Phase I – Lit Review prepared for US Army Medical Res Develop Com by SRI Int, Menlo Park, CA. Contract No DAMD17-78-C-8081 AD-A082372
6. *Urbanski T* (1985) Chemistry and technology of explosives, vol 3, Pergamon Press, NY, p 22
7. *Roth J* (1978) [Nitration. VI. Technology of nitration], TNT, in: Encyclopedia of explosives and related items, vol 8, Kaye S M (ed) Dover, NJ, Large Caliber Weapon Systems Lab, US Army Armament Res Develop Com AD-A057762, p N48
8. *Urbanski T* (1985) Chemistry and technology of explosives, vol 4, Pergamon Press, New York, p 151
9. *Clarke H T, Hartman W W* (1941) 1,3,5-Trinitrobenzene, in: Organic syntheses, Coll vol I, Gilman H, Blatt A H (eds) Wiley, New York, p 541
10. *Urbanski T* (1985) Chemistry and technology of explosives, vol 1, Pergamon Press, New York, p 242

11. *Fedoroff B T, Sheffield O E* (eds) (1966) Cyclotrimethylenetrinitramine, cyclonite or RDX, in: Encyclopedia of explosives and related items, vol 3, Dover, NJ, Picatinny Arsenal AD-653029, p C611
12. *Urbanski T* (1985) Chemistry and technology of explosives, vol 3, Pergamon Press, NY, p 87 + 111
13. *Fedoroff B T, Sheffield O E,* (eds) (1966) Cyclotetramethylenetetranitramine, 1,3,5,7-tetranitro-1,3,5,7-tetrazacyclooctane; octahydro-1,3,5,7-tetranitro-1,3,5,7-tetrazine; homocyclonite; octogen or HMX, in: Encyclopedia of explosives and related items, vol 3, Dover, NJ, Picatinny Arsenal AD-653029, p C605
14. *Urbanski T* (1985) Chemistry and technology of explosives, vol 3, Pergamon Press, NY, p 117
15. *Bedford C D, Deas B D, Broussard M M, Geigel M A, Marynowski C W* (1981) Preparation and purification of multigram quantities of TAX and SEX, Third Phase final report prepared for US Army Med Res Develop Com by SRI Int, Menlo Park, CA. Contr No DAMD17-80-C-0013 AD-A122816
16. *Bedford C D* (1983) Preparation and purification of multigram quantities of SEX & TAX, Phase IV final report prepared for US Army Med Res Develop Com by SRI Int, Menlo Park, CA. Contr No DAMD17-80-C-2092 AD-A146377
17. *Urbanski T* (1985) Chemistry and technology of explosives, vol 3, Pergamon Press, NY, p 40
18. *Urbanski T* (1985) Chemistry and technology of explosives, vol 1, Pergamon Press, New York, p 472
19. *Urbanski T* (1985) Chemistry and technology of explosives, vol 2, Pergamon Press, New York, p 175
20. *Urbanski T* (1985) Chemistry and technology of explosives, vol 2, Pergamon Press, New York, p 32
21. *Urbanski T* (1985) Chemistry and technology of explosives, vol 2, Pergamon Press, New York, p 142
22. *Urbanski T* (1985) Chemistry and technology of explosives, vol 2, Pergamon Press, New York, p 149
23. *Urbanski T* (1985) Chemistry and technology of explosives, vol 2, Pergamon Press, New York, p 157
24. *Conklin C, Pristera F* (1958) Preparation and physical properties of di- and trinitrotoluene isomers. Techn report 2525, Picatinny Arsenal, Dover, NJ. AD-200206
25. *Dennis W H, Rosenblatt D H, Blucher W G, Coon C L* (1975) J Chem Eng Data 20: 202
26. *Gilbert E E, Leccacorvi J R* (1976) Propellants Explos 1: 89
27. *Kaye S M* (ed) (1980) Sweating, in: Encyclopedia of explosives and related items, vol 9, Dover, NJ, Large Caliber Weapon Systems Lab, US Army Armament Res Develop Com AD-A097595, p 5256
28. *Urbanski T* (1985) Chemistry and technology of explosives, vol 1, Pergamon Press, New York, p 571
29. *Capellos C, Fisco W J, Ribaudo C, Hogan V D, Campisi J, Murphy F X, Castorina T C, Rosenblatt D H* (1979) Kinetic studies and product characterization during the basic hydrolysis of glyceryl nitrate esters. Techn report ARLCD-TR-79022, Large Caliber Weapon Systems Lab, US Army Armament Res Develop Com, Dover, NJ. AD-A075340
30. *Dunstan I, Griffiths J V, Harvey S A* (1965) J Chem Soc 1319
31. *Nichols P L Jr, Magnusson A B, Ingham* (1953) J Am Chem Soc 75: 4255
32. *Urbanski T* (1985) Chemistry and technology of explosives, vol 2, Pergamon Press, New York, p 127
33. *Fauth M I* (1986) An assessment of the environmental effects of Otto Fuel II combustion products during underwater operations. Ordnance Env Supp Office, Naval Ordnance Station, Indian Head, MD
34. *Grain C F* (1982) Vapor Pressure, in: Handbook of chemical property estimation methods: Environmental behavior of organic compounds, Lyman W J, Reehl W F, Rosenblatt D H, (eds) chap 14, McGraw-Hill, New York
35. *Yalkowsky S,* (1979) Ind Eng Chem Fundam 18: 108
36. *Urbanski T* (1985) Chemistry and technology of explosives, vol 1, Pergamon Press, New York, p 249
37. *Urbanski T* (1985) Chemistry and technology of explosives, vol 1, Pergamon Press, New York, p 294

38. *Fedoroff B T, Sheffield O E* (eds) (1974) Ethyleneglycol dinitrate (EGDN or EGcDN), in: Encyclopedia of explosives and related items, vol 6, Dover, NJ, Picatinny Arsenal. AD-A011845, p E259

39. *Fedoroff B T, Sheffield O E* (eds) (1974) Nitroguanidine, in: Encyclopedia of explosives and related items, vol 6, Dover, NJ, Picatinny Arsenal. AD-A011845, p G154

40. *Roth J* (1978) Picric acid, in: Encyclopedia of explosives and related items, vol 8, Kaye S M (ed) Dover, NJ, Large Caliber Weapon Systems Lab, US Army Armament Res Develop Com AD-A057762, p P285

41. *Lyman WJ* (1982) Solubility in water, in: Handbook of chemical property estimation methods: Environmental behavior of organic compounds, Lyman W J, Reehl W F, Rosenblatt D H, (eds) chap 2, McGraw-Hill, New York

42. *Banerjee S, Yalkowsky S H, Valvani S* (1980) Environ Sci Technol 14: 1227

43. *Cundall R B, Palmer T F, Wood C E C* (1981) J Chem Soc Faraday Trans 177: 711

44. *Pella P A* (1977) J Chem Thermodyn 9: 301

45. *Maksimov Yu Ya* (1963) Teoriya Vzryvchatykh Veshchestv, Sb. Statei 1963: 546, CA 59: 12208f; (1968) Zh Fiz Khim 42: 2921, CA 70: 61315y

46. *Urbanski T* (1985) Chemistry and technology of explosives, vol 3, Pergamon Press, NY

47. *Crater W deC* (1929) Ind Eng Chem 21: 674

48. *Weast R C, Astle M J* (ed) (1979) Physical constants of organic compounds, CRC handbook of chemistry and physics, 60th ed, Boca Raton, FL

49. *Lyman W J, Reehl W F, Rosenblatt D H* (eds) (1982) Handbook of chemical property estimation methods: Environmental behavior of organic compounds, New York, McGraw-Hill

50. *Major M A* (1989) U.S. Army Biomed Res Develop Lab, Ft Detrick, Frederick, MD (unpublished data)

51. *Jenkins T F* (1989) Development of an analytical method for the determination of extractable nitroaromatics and nitramines in soils. Ph D thesis, Univ of New Hampshire, Durham, N H

52. *Hansch C, Leo A* (1979) Substituent constants for correlation analysis in chemistry and biology, Wiley, New York,

53. *Lyman W J* (1982) Adsorption coefficient for soils and sediments, in: Handbook of chemical property estimation methods: Environmental behavior of organic compounds, Lyman W J, Reehl W F, Rosenblatt D H, (eds) chap 4, McGraw-Hill, New York

54. *Lyman W J, Loreti C P* (1987) Prediction of soil and sediment sorption for organic compounds. Final report prepared for Monitoring and Data Support Div, Office of Water Regul and Standards, US Env Prot Agency by Arthur D Little, Inc, Cambridge, MA, Contr No 68-01-6951, Task 15

55. *Bysshe S E* (1982) Bioconcentration factor, in: Handbook of chemical property estimation methods: Environmental behavior of organic compounds, Lyman W J, Reehl W F, Rosenblatt D H, (eds) chap 5, McGraw-Hill, New York

56. *Isnard P, Lambert S* (1988) Chemosphere 17: 21

57. *Hartley W R* (1981) Evaluation of selected subacute effects of 2,4-dinitrotoluene on the bluegill sunfish. Ph D thesis, Tulane Univ, New Orleans, LA

58. US Env Prot Agency, Office of Water Planning and Standards, Criteria and Standards Div (1980) Ambient water quality criteria for dinitrotoluene, EPA 440/5-80-045, Wash, DC. NTIS PB 296794

59. *Kenaga E E* (1980) Environ Sci Technol 14: 553

60. *Garten C T Jr, Trabalka J R* (1983) Environ Sci Technol 17: 590

61. *Travis C C, Arms A D* (1988) Environ Sci Technol 22: 271

62. *Thonney M L, Oberbauer A M, Duhaime D J, Jenkins T C, Firth N L* (1984) J Nutr 114: 1777

63. *Lewis R J, Sweet D V* (1984) Registry of toxic effects of chemical substances, 1983 Suppl to the 1981–82 edn. US Dept of Health and Human Services, Nat Inst Occup Safety and Health, Cincinnati, OH

64. *Rappoport Z* (ed), (1977) CRC Handbook of tables for organic compound identification, 3rd ed, Table XX, CRC Press, Cleveland, OH

65. *Kaye S M* (ed) (1980) Sym trinitrobenzene, in: Encyclopedia of explosives and related items, vol 9, Dover, NJ, Large Caliber Weapon Systems Lab, US Army Armament Res Develop Com AD-A097595, p T375

66. *Rosenblatt D H* (1986) Contaminated soil cleanup objectives for Cornhusker Army Ammunition Plant. Techn report 8603, US Army Med Bioeng Res Develop Lab, Fort Detrick, Frederick, MD AD-A1669179

67. *Urbanski T* (1985) Chemistry and technology of explosives, vol 1, Pergamon Press, New York, p 259
68. *Hall P G* (1971) Trans Faraday Soc 67(1): 556
69. *Kishore K* (1977) Propellants Explos 2: 78
70. *Leiga A G, Sarmousakis J N* (1966) J Phys Chem 70: 3544
71. *Glover D J, Hoffsommer J C* (1973) Bull Environ Contam Toxicol 10: 302
72. *Merrill E J* (1965) J Pharm Sci 54: 1670
73. US Army Materiel Command (1971) Properties of explosives of military interest. Engineering design handbook. Explosives Series, AMC Pamphlet 706-177, Wash DC
74. *Spanggord R J, Chou T W, Mill T, Podoll R T, Harper J C, Tse D S* (1985) Environmental fate of nitroguanidine, diethylene glycol dinitrate, and hexachloroethane smoke. Final report, Phase I, prepared for US Army Med Res Develop Com by SRI Int, Menlo Park, CA. Contr No DAMD17-84-C-4252 AD-A185050
75. *Roberts M S, Cossum P A, Kowaluk E A, Polack A E* (1983) Int J Pharm 17: 145
76. *Spanggord R J, Mill T, Chou T W, Mabey W R, Smith J H, Lee S* (1980) Environmental fate studies on certain munition wastewater constituents. Final report, Phase II-Lab studies prepared for US Army Med Res Develop Com by SRI Int, Menlo Park, CA. Contr No DAMD17-78-C-8081 AD-A099256
77. *Bentley R E, Dean J W, Ells S J, LeBlanc G A, Sauter S, Buxton K S, Sleight B H III* (1978) Laboratory evaluation of the toxicity of nitroglycerine to aquatic organisms. Final report prepared for US Army Med Res Develop Com by EG&G Bionomics, Wareham, MA. Contr No DAMD17-74-C-4101, AD-A061739
78. *Leggett D C* (1977) J Chromatogr 133: 83
79. *Cooper K R, Burton D T, Goodfellow W L, Rosenblatt D H* (1984) J Toxicol Environ Health 14: 731
80. *Yinon J* (1977) Analysis of explosives, in: Critical reviews in analytical chemistry Campbell B (ed) CRC Press, Boca Raton, p 1
81. *Vouros P, Petersen B A, Colwell L, Karger B L, Harris H* (1977) Anal Chem 49: 1039
82. *Burrows E P, Brueggemann E E* (1985) J Chromatogr 329: 285
83. *Bauer C F, Grant C L, Jenkins T F* (1986) Anal Chem 58: 176
84. *Jenkins T F, Leggett D C, Grant C L, Bauer C F* (1986) Anal Chem 58: 170
85. *Jenkins T F, Grant C L* (1987) Anal Chem 59: 1326
86. *Kaplan D L, Kaplan A M* (1982) Anal Chim Acta 136: 425
87. *Murphy L J, Siggia S, Uden P C* (1986) J Chromatogr 366: 161
88. *Burrows E P, Brueggemann E E, Hoke S H, McNamee E H, Baxter L J* (1984) Nitroguanidine wastewater pollution control technology: Phase II. Wastewater characterization and analytical methods development for organics. Techn report 8311, US Army Med Bioeng Res Develop Lab, Fort Detrick, Frederick, MD. AD-A141176
89. *Bratin K, Kissinger P T, Briner R C, Bruntlett C S* (1981) Anal Chim Acta 130: 295
90. *Krull I S, Ding X D, Selavka C, Bratin K, Forcier G* (1984) J Forensic Sci 29: 449
91. *Lloyd J B F* (1983) J Chromatogr 257: 227
92. *Maskarinec M P, Manning D L, Harvey R W, Griest W H, Tomkins B A* (1984) J Chromatogr 302: 51
93. *Manning D L, Maskarinec M P* (1987) Analysis of nitroguanidine in aqueous solutions by HPLC with electrochemical detection and voltammetry. Final report prepared for US Army Toxic and Haz Materials Agency by Oak Ridge Natl Lab, Oak Ridge, TN. AD-A179494
94. *Lafleur A L, Morriseau B D* (1980) Anal Chem 52: 1313
95. *Selavka C M, Tontarski R E, Jr, Strobel R A* (1987) J Forensic Sci 32: 941
96. *Yinon J, Hwang D G* (1983) J Chromatogr 268: 45
97. *Berberich D W, Yost R A, Fetterolf D D* (1988) J Forensic Sci 33: 946
98. *Voyksner R D, Yinon J* (1986) J Chromatogr 354: 393
99. *Douse J M F* (1981) J Chromatogr 208: 83
100. *Douse J M F* (1982) J Chromatogr 234: 415
101. *Glover D J, Hoffsommer J C, Kubose D A* (1977) Anal Chim Acta 88: 381
102. *Sioufi A, Pommier F* (1985) J Chromatogr 339: 117
103. *Douse J M F* (1985) J Chromatogr 328: 155
104. *Richard J J, Junk G A* (1986) Anal Chem 58: 725
105. *Pereira W E, Short D L, Manigold D B, Roscio P K* (1979) Bull Environ Contam Toxicol 21: 554
106. *Tamiri T, Zitrin S* (1986) J Energ Mater 4: 215

107. *Zitrin S* (1982) Org Mass Spectrom 17:74
108. *Yinon J* (1980) J Forensic Sci 25:401
109. *Gielsdorf W* (1981) Fresenius Z Anal Chem 308:123
110. *Schulten H R, Lehmann W D* (1977) Anal Chim Acta 93:19
111. *McLafferty F W* (ed) (1983) Tandem mass spectrometry, Wiley, New York
112. *Carper W R, Dorey R C, Tomer K B, Crow F W* (1984) Org Mass Spectrom 19:623
113. *Yinon J* (1987) Org Mass Spectrom 22:501
114. *Yinon J, Harvan D J, Hass J R* (1982) Org Mass Spectrom 17:321
115. *McLuckey S A, Glish G L, Carter J A* (1985) J Forensic Sci 30:773
116. *McLuckey S A, Glish G L, Kelley P E* (1987) Anal Chem 59:1670
117. *Yinon J* (1987) Mass spectrometry of explosives in: Yinon J (ed) Forensic Mass Spectrometry, CRC Press, Boca Raton, FL
118. *Bagnato L, Grasso G* (1986) J Chromatogr 357:440
119. *Carlson M, Thompson R D* (1986) J Chromatogr 368:472
120. *Ryon M G, Pal B C, Talmage S S, Ross R H* (1984) Database assessment of the health and environmental effects of munition production waste products. Final report prepared for US Army Med Res Develop Com by Oak Ridge Natl Lab, Oak Ridge, TN, ORNL-6018
121. *Burrows W D, Paulson E T, Carnahan R P* (1989) Biological treatment of composition B wastewaters. II. Analysis of performance of Holston Army Ammunition Plant wastewater treatment facility, Jan 1985 through Aug 1986. Techn report 8806, US Army Biomed Res Develop Lab, Fort Detrick, Frederick, MD
122. *Dacre J, Rosenblatt D H* (1974) Mammalian toxicology and toxicity to aquatic organisms of four important types of waterborne munitions pollutants – An extensive literature evaluation. Techn report 7403, US Army Biomed Res Develop Lab, Aberdeen Proving Ground, MD. AD-778725
123. *Spanggord R J, Gibson B W, Keck R G, Thomas D W, Barkley J J* (1982) Environ Sci Technol 16:229
124. *Sanoki S L, Simon P B, Weitzel R L, Jerger D E, Schenk J E* (1976) Aquatic field surveys at Iowa, Radford, and Joliet Army Ammunition Plants, vol I. Iowa Army Ammunition Plant. Final report prepared for US Army Med Res Develop Com by Environ Control Techn Corp, Ann Arbor, MI. Contr No DAMD17-75-C-5046. AD-A036776
125. *Envirodyne Engineers, Inc.*: Milan Army Ammunition Plant Contamination Survey. Final report prepared for U.S. Army Toxic & Hazardous Materials Agency as Report No. DRXTH-FS-80-050, Envirodyne Engineers, Inc. St. Louis, MO Contract No. J-79-569, 1980. AD-B053362L
126. *Hale V Q, Stanford T B, Taft L G* (1979) Evaluation of the environmental fate of munitions compounds in soil. Final report prepared for US Army Med Res Develop Com by Battelle Columbus Labs, Columbus, O H. Contr No DAMD17-76-C-6065. AD-A082874
127. *Osmon J L, Andrews C C* (1978) The biodegradation of TNT in enhanced soil and compost systems. Techn report 77032, US Army Armament Res Develop Com, Dover, NJ. AD-A054375
128. *Kayser E G, Burlinson N E, Rosenblatt D H* (1984) Kinetics of hydrolysis and products of hydrolysis and photolysis of tetryl. Techn report 84-78, Naval Surface Weapons Center, Silver Spring, MD. AD-A153144
129. *Amerkhanova N N, Naumova R P* (1978) Mikrobiologiya 47:393; CA 89:125844x
130. *Chambers C W, Tabak H H, Kabler P W* (1963) J Water Pollut Control Fed 35:1517
131. *McCormick N G, Feeherry F E, Levinson H S* (1976) Appl Environ Microbiol 31:949
132. *Nay M W Jr, Randall C W, King P H* (1974) J Water Pollut Control Fed 46:485
133. *Osmon J L, Klausmeier R E* (1972) Dev Ind Microbiol 14:247
134. *Parrish F W* (1977) Appl Environ Microbiol 34:232
135. *Traxler R W, Wood E, Delaney J M* (1974) Dev Ind Microbiol 16:71
136. *Weitzel R L, Simon P B, Jerger D E, Schenk J E*: Aquatic Field Survey at Iowa Army Ammunition Plant. Final report prepared for U.S. Army Medical Research and Development Command by Environmental Control Technology Corp., Ann Arbor, MI. Contract No. DAMD17-74-C-4124, 1975. AD-A014300
137. *Enzinger R M* (1971) Special study of the effect of α-TNT on microbiological systems and the determination of the biodegradability of α-TNT. Report, US Army Environ Health Agency, Aberdeen Prov Ground, MD. AD-738497
138. *Won W D, Heckly R J, Glover D J, Hoffsommer J C* (1974) Appl Microbiol 27:513

139. *Carpenter D F, McCormick N G, Cornell J H, Kaplan A M* (1978) Appl Environ. Microbiol. 35: 949
140. *Kaplan D L, Kaplan A M* (1982) Appl Environ Microbiol 44: 757
141. *Isbister J D, Anspach G L, Kitchens J F, Doyle R C* (1984) Microbiologica 7: 47
142. *Bell B A, Burrows W D, Sotsky L, Carraza J A* (1984) Munitions wastewater treatment in semicontinuous activated sludge treatment systems. Report ARLCD-CR-84029, US Army Armament Res Develop Cent, Large Caliber Weapon Systems Lab, Dover, NJ. AD-B087493
143. *Mitchell W R, Dennis W H Jr* (1982) J Environ Sci Health A17: 837
144. *Mitchell W R, Dennis W H Jr, Burrows E P* (1982) Microbial interactions with several munitions compounds: 1,3-Dinitrobenzene, 1,3,5-trinitrobenzene, and 3,5-dinitroaniline. Techn report 8201, US Army Med Bioeng Res Develop Lab, Fort Detrick, Frederick, MD. AD-A116651
145. *McCormick N G, Cornell J H, Kaplan A M* (1978) Appl Environ Microbiol 35: 945
146. *Spanggord R J, Mabey W R, Mill T, Chou T W, Smith J H, Lee S* (1981) Environmetal fate studies on certain munition wastewater constituents. Final report, Phase III, Part II–Lab Studies, prepared for US Army Med Res Develop Com by SRI Int, Menlo Park, CA. Contr No DAMD17-78-C-8081. AD-A133987
147. *Liu D, Thomson K, Anderson A C* (1984) Appl Environ Microbiol 47: 1295
148. *Jerger D E, Simon P B, Weitzel R L, Schenk J E* (1976) Aquatic field surveys at Iowa, Radford, and Joliet Army Ammunition Plants. vol III. Microbiological investigation, Iowa and Joliet Army Ammunition Plants. Final report prepared for US Army Med Res Develop Com by Environ Control Techn Corp, Ann Arbor, MI. Contract No DAMD17-75-C-5046. AD-A036778
149. *Keirn M A, Stratton C L, Mousa J J, Donds J D, Winegardner D L, Prentice H S, Adams W D, Powell J J* (1981) Environmental survey of Alabama Army Develop Com by Environ Sci and Eng, Inc, Gainsville, FL. Contr No DAAK11-79-C-0131. AD-B059961
150. *Spanggord R J, Mabey W R, Chou T W, Lee S, Alferness P L, Tse D S, Mill T* (1983) Environmental fate studies of HMX. Final report, Phase II - Detailed studies, prepared for U.S. Army Med Res Develop Com by SRI Int, Menlo Park, CA. Contr No DAMD17-82-C-2100. AD-A145122
151. *Spanggord R J, Mabey W R, Mill T, Chou T W, Smith J H, Lee S, Roberts D* (1983) Environmental fate studies on certain munition wastewater constituents. Final report, Phase IV - Lagoon model studies, prepared for US Army Med Res Develop Com by SRI Int, Menlo Park, CA. Contr No DAMD17-78-C-8081. AD-A138550
152. *McCormick N G, Cornell J H, Kaplan A M* (1981) Appl Environ Microbiol 42: 817
153. *McCormick N G, Cornell J H, Kaplan A M* (1984) The anaerobic biotransformation of RDX, HMX, and their acetylated derivatives. Techn report 85/007, US Army Natick Res Develop Labs, Natick, MA. AD-A149464
154. *McCormick N G, Cornell J H, Kaplan A M* (1984) The fate of hexahydro-1,3,5-trinitro-1,3,5-triazine (RDX) and related compounds in anaerobic denitrifying continuous culture systems using simulated wastewater. Techn report 85/008, US Army Natick Res Develop Labs, Natick, MA. AD-A149462
155. *Kaplan D L, Cornell J H, Kaplan A M* (1982) Environ Sci Technol 16: 488
156. *Kaplan D L, Kaplan A M* (1985) Degradation of nitroguanidine in soils. Techn report 85/047, US Army Natick Res Develop Cent, Natick, MA. AD-A157859
157. *Wendt T M, Cornell J H, Kaplan A M* (1978) Appl Environ Microbiol 36: 693
158. *Cornell J H, Wendt T M, McCormick N G, Kaplan D L, Kaplan A M* (1981) Biodegradation of nitrate esters used as military propellants — A status report. Techn report 81/029, US Army Natick Res Develop Labs, Natick, MA. AD-A149665
159. *Kaplan D L, Walsh J T, Kaplan A M* (1982) Environ Sci Technol 16: 723
160. *Burlinson N E, Kaplan L A, Adams C E* (1973) Photochemistry of TNT: Investigation of the "Pink Water" problem. Techn report 73-172, Naval Ordinance Lab, Silver Spring, MD. AD-767670
161. *Burlinson N E, Sitzmann M E, Glover D J, Kaplan L A* (1979) Photochemistry of TNT and Related Nitroaromatics: Part III. Techn report 78-198, Naval Surface Weapons Center, Silver Spring, MD. AD-B045845
162. *Kaplan L A, Burlinson N E, Sitzmann M E* (1975) Photochemistry of TNT: Investigation of the "Pink Water" problem, Part II. Techn report 75-152, Naval Surface Weapons Center, Silver Spring, MD. AD-A020072

163. *Burlinson N E* (1980) Fate of TNT in an aquatic environment: Photodecomposition vs. biotransformation. Techn report 79-445, Naval Surface Weapons Center, Silver Spring, MD. AD-B045846
164. *Ho P C* (1986) Environ Sci Technol 20: 260
165. *Burlinson N E, Sitzmann M E, Kaplan L A, Kayser E* (1979) J Org Chem 44: 3695
166. *Mabey W R, Tse D, Baraze A, Mill T* (1983) Chemosphere 12: 3
167. *Simmons M S, Zepp R G* (1986) Water Research 20: 899
168. *Zepp R G, Schlotzhauer P F, Sink R M* (1985) Environ Sci Technol 19: 74
169. *Burrows W D, Brueggemann E E* (1986) Tertiary treatment of effluent from Holston AAP industrial liquid waste treatment facility. V. Degradation of nitramines in Holston AAP wastewaters by ultraviolet radiation. Techn report 8602, US Army Med Bioeng Res Develop Lab, Fort Detrick, Frederick, MD. AD-A176195
170. *Glover D J, Hoffsommer J C* (1979) Photolysis of RDX in aqueous solution, with and without ozone. Techn report 78-175, Naval Surface Weapons Center, Silver Spring, MD. AD-080195
171. *Burrows W D* (1983) Tertiary treatment of effluent from Holston AAP industrial liquid waste treatment facility. III. Ultraviolet radiation and ozone studies: TNT, RDX, HMX, TAX, and SEX. Techn report 8306, US Army Med Bioeng Res Develop Lab, Fort Detrick, Frederick, MD. AD-A137672
172. *Glover D J, Hoffsommer J C* (1979) Photolysis of RDX, Identification and reactions of products. Techn report 79-349, Naval Surface Weapons Center, Silver Spring, MD. AD-B047443
173. *Kubose D A, Hoffsommer J C* (1977) Photolysis of RDX in aqueous solution. Initial studies. Techn report 77-20, Naval Surface Weapons Center, Silver Spring, MD. AD-A042199
174. *Burrows W D, Schmidt M O, Chyrek R H, Noss C I* (1988) Photochemistry of aqueous nitroguanidine. Techn report 8808, US Army Med Bioeng Res Develop Lab, Fort Detrick, Frederick, MD. AD-A203200
175. *Capellos C, Fisco W J, Ribaudo C, Hogan V D, Campisi J, Murphy F X, Castorina T C, Rosenblatt D H* (1982) Int J Chem Kinet 14: 903
176. *Capellos C, Fisco W J, Ribaudo C, Hogan V D, Campisi J, Murphy F X, Castorina T C, Rosenblatt D H* (1984) Int J Chem Kinet 16: 1009
177. *Capellos C, Fisco W J, Ribaudo C, Hogan V D, Campisi J, Murphy F X, Castorina T C, Rosenblatt D H* (1984) Int J Chem Kinet 16: 1027
178. *Kaplan D L, Cornell J H, Kaplan A M* (1981) Decomposition of the epoxides glycidol and glycidyl nitrate. Techn report 81/018, US Army Natick Res Develop Labs, Natick, MA. AD-A101630
179. *Yinon J, Hwang D G* (1986) J Energ Mater 4: 305
180. *Yinon J, Hwang D G* (1987) J Chromatogr 394: 253
181. *Palazzo A J, Leggett D C* (1986) J Environ Qual 15: 49
182. *El-hawari A M, Hodgson J R, Winston J M, Sawyer M D, Hainje M, Lee C C* (1981) Species differences in the disposition and metabolism of 2,4,6-trinitrotoluene as a function of route of administration. Final report, prepared for US Army Med Res Develop Com by Midwest Res Inst, Kansas City, MO. Contr No DAMD17-76-C-6066. AD-A114025
183. *Langseth-Manrique K, Bredesen J E, Greibrokk T* (1986) J High Resolut Chromatogr Commun 9: 643
184. *Sioufi A, Pommier F, DuBois J P* (1987) J Chromatogr 413: 101
185. *Svobodova X, Kovacova D, Ostrovska V, Pechova A, Polacikova O, Kusala S, Svoboda M* (1988) J Chromatogr 425: 391
186. *Lloyd J B F* (1986) J Forensic Sci Soc 26: 341
187. *Long R M, Rickert D E* (1982) Drug Metab Dispos 10: 455
188. *Rickert D E, Long R M* (1981) Drug Metab Dispos 9: 226
189. *Levine R J, Turner M J, Crume Y S, Dale M E, Starr T B, Rickert D E* (1985) J Occupat Med 27: 627
190. *Turner M J Jr, Levine R J, Nystrom D D, Crume Y S, Rickert D E* (1985) Toxicol Appl Pharmacol 80: 166
191. *Schneider N R, Bradley S L, Andersen M E* (1977) Toxicol Appl Pharmacol 39: 531
192. *Schneider N R, Bradley S L, Andersen M E* (1978) Toxicol Appl Pharmacol 46: 163
193. *Woody R C, Kearns G L, Brewster M A, Turley C P, Sharp G B, Lake R S* (1986) Clin Toxicol 24: 305
194. *Turley C P, Brewster M A* (1987) J Chromatogr 421: 430

195. *Ho B, Tillotson J A, Kincannon L C, Simboli P B, Korte D W Jr* (1988) Fund Appl Toxicol 10:453
196. *Stokinger H E* (1982) Aliphatic nitro compounds, nitrates, nitrites, in: Patty's industrial hygiene and toxicology, 3rd ed, vol 2C, Clayton G D, Clayton F E (eds) Wiley, New York, p 4169
197. *Forman S A, Helmkamp J C, Bone C M* (1987) J Occupat Med 29:445
198. *Horvath E P, Ilka R A, Boyd J, Markham T* (1981) Am J Indust Med 2:365
199. *Beard R R, Noe J T* (1981) Aromatic nitro and amino compounds, in: Patty's industrial hygiene and toxicology, 3rd ed, vol 2A, Clayton G D, Clayton F E (eds) Wiley, New York, p 2413
200. *Rosenblatt D H* (1980) Toxicology of explosives and propellants, in: Encyclopedia of explosives and related items, vol 9, Kaye S M (ed) Dover, NJ, Large Caliber Weapon Systems Lab, US Army Armament Res Develop Com AD-A097595, p T332
201. *Etnier E L* (1987) Water quality criteria for hexahydro-1,3,5-trinitro-1,3,5-triazine (RDX). Final report prepared for US Army Med Res Develop Com by Oak Ridge Natl Lab, Oak Ridge, TN, ORNL-6178
202. *McLellan W, Hartley W R, Brower M* (1988) Health advisory for hexahydro-1,3,5-trinitro-1,3,5-triazine. Office of Drinking Water, US Env Prot Agency, Wash, DC
203. *Kaplan A S, Berghout C F, Peczenik A* (1965) Arch Environ Health 10:877
204. *Ketel W B, Hughes J R* (1972) Neurology 22:871
205. *Stone W J, Paletta T L, Heiman E M, Bruce J I, Knepshield J H* (1969) Arch Intern Med 124:726
206. *Hathaway J A, Buck C R* (1977) J Occupat Med 19:269
207. *McLellan W, Hartley W R, Brower M* (1988) Health advisory for octahydro-1,3,5,7-tetranitro-1,3,5-triazine. Office of Drinking Water, US Env Prot Agency, Wash, DC
208. *Brown L D, Wheeler C R, Korte D W Jr* (1988) Acute oral toxicity of nitro-guanidine in male and female rats. Inst report 264, Letterman Army Inst of Res, Presidio of San Fran, CA. AD-A192694
209. *Hiatt G F S, Sano S K, Wheeler C R, Korte D W Jr* (1988) Acute oral toxicity of nitroguanidine in mice. Inst report 265, Letterman Army Inst of Res, Presidio of San Fran, CA. AD-A194952
210. *Morgan E W, Pearce M J, Zaucha G M, Lewis C M, Makovec G T, Korte D W Jr* (1988) Ninety-day subchronic oral toxicity study of nitroguanidine in rats. Inst report 306, Letterman Army Inst of Res, Presidio of San Fran, CA
211. *Leonard T B, Adams T, Popp J A* (1986) Carcinogenesis 7:1797
212. *Levine R J, Andjelkovich D A, Kersteter S L, Arp E W, Balogh S A, Blunden P B, Stanley J M* (1986) J Occupat Med 28:811
213. *Hogstedt C, Andersson K* (1979) J Occupat Med 21:553
214. *Ryon M G* (1987) Water quality criteria for 2,4,6-trinitrotoluene. Final report prepared for US Army Med Res Develop Com by Oak Ridge Natl Lab, Oak Ridge, TN, ORNL-6304
215. *Etnier E L* (1987) Water quality criteria for 2,4-dinitrotoluene and 2,6-dinitrotoluene. Final report prepared for US Army Med Res Develop Com by Oak Ridge Natl Lab, Oak Ridge, TN, ORNL-6312
216. *Smith J G* (1986) Water quality criteria for nitroglycerin. Final report prepared for US Army Med Res Develop Com by Oak Ridge Natl Lab, Oak Ridge, TN, ORNL-6180
217. *Cuthbert J A, D'Arcy-Burt K J, Carr S M A* (1985) HMX: Acute toxicity tests in laboratory animals. Final report prepared for US Army Med Res Develop Com by Inveresk Res Int, Ltd, Musselburgh, Scotland. Contr No DAMD17-80-C-0053. AD-A171598
218. *Frost D F, Brown L D, Morgan E W, Korte D W* (1988) Acute toxicity of DEGDN and two DEGDN-containing solid propellants, DIGL-RP and JA-2. Proc 1988 JANNAF Safety and Env Prot Subcommittee Meeting, Monterey, CA
219. *Levine B S, Furedi E M, Gordon D E, Lish P M, Barkley J J* (1984) Toxicology 32:253
220. *Dilley J V, Tyson C A, Newell G W* (1978) Mammalian toxicological evaluation of TNT wastewaters. vol 2: Acute and subacute mammalian toxicity of TNT and the LAP mixture. Final report prepared for US Army Med Res Develop Com by SRI Int, Menlo Park, CA. Contr No DAMD17-76-C-6050 AD-A081590
221. *Levine B S, Rust J H, Burns J M, Lish P M* (1983) Determination of the chronic mammalian effects of TNT. Twenty-six week subchronic oral toxicity study of trinitrotoluene in the beagle dog. Final report, Phase II, prepared for US Army Med Res Develop Com by IIT Res Inst, Chicago, IL. Contr No DAMD17-79-C-9120. AD-A157082
222. *Furedi E M, Levine B S, Gordon D E, Rac V S, Lish P M* (1984) Determination of the chronic mammalian toxicological effects of TNT (Twenty-four month chronic toxicity/carcinogenicity

study of trinitrotoluene in the Fischer 344 rat). Report prepared for US Army Med Res Develop Com by IIT Res Inst, Chicago, IL. Contr No DAMD17-79-C-9120. AD-A168637

223. *Ellis H V, Hagenson J H, Hodgson J R, Minor J L, Hong C B, Ellis E R, Girvin J D, Helton D O, Herndon B L, Lee C C* (1979) Mammalian toxicology of munitions compounds. Phase III: Effects of lifetime exposure. Part I: 2,4-Dinitrotoluene. Report prepared for US Army Med Res Develop Com by Midwest Res Inst, Kansas City, MO. Contr No DAMD17-74-C-4073. AD-A077692

224. *Hong C B, Ellis H V, Lee C C, Sprinz H, Dacre J C, Glennon J P* (1985) J Am Coll Toxicol 4: 257

225. *Lee C C, Hong C B, Ellis H V, Dacre J C, Glennon J P* (1985) J Am Coll Toxicol 4: 243

226. *Normandy J, Gordon L, Hartley W R,* (1987) Trinitroglycerol health advisory. Office of Drinking Water, US Env Prot Agency, Washington, DC

227. *Ellis H V, Hong C B, Lee C C, Dacre J C, Glennon J P* (1984) Fundam Appl Toxicol 4: 248

228. *Levine B S, Furedi E M, Rac V S, Gordon D E, Lish P M* (1983) Determination of the chronic mammalian toxicological effects of hexahydro-1,3,5-trinitro-1,3,5-triazine. Twenty-four month chronic toxicity/carcinogenicity study of RDX in the Fischer 344 rat. Final report Phase V, Vol. 1 prepared for US Army Med Res Develop Com by IIT Res Inst, Chicago, IL. Contr No DAMD17-79-C-9161. AD-A160744

229. *Lish P M, Levine B S, Furedi E M, Sagartz J M, Rac V S* (1984) Determination of the chronic mammalian toxicological effects of hexahydro-1,3,5-trinitro-1,3,5-triazine (RDX). Twenty-four month chronic toxicity/carcinogenicity study of RDX in the B6C3F1 hybrid mouse. Final report, Phase VI, Vol. 1 prepared for US Army Med Res Develop Com by IIT Res Inst, Chicago, IL. Contr No DAMD17-79-C-9161. AD-A181766

230. *van Esch G J, Vink H H, van Genderen H* (1957) Nature 180: 509

231. *Cody T E, Witherup S, Hastings L, Stemmer K, Christian R T* (1981) J Toxicol Environ Health 7: 829

232. *Linder R E, Hess R A, Strader L F* (1986) J Toxicol Environ Health 19: 477

233. *Liu D H W, Spanggord R.J, Bailey H C, Javitz, H S, Jones, D C L* (1984) Toxicity of TNT wastewaters to aquatic organisms, vol II. Acute toxicity of condensate wastewater and 2,4-dinitrotoluene. Final report prepared for US Army Med Res Develop Com by SRI Int, Menlo Park, CA Contr No DAMD17-75-C-5056. AD-A142145

234. *Liu D H W, Spanggord R J, Bailey H C, Javitz, H S, Jones, D C L* (1984) Toxicity of TNT wastewaters to aquatic organisms, vol I. Acute toxicity of LAP wastewater and 2,4,6-trinitro-toluene. Final report prepared for US Army Med Res Develop Com by SRI Int, Menlo Park, CA Contr No DAMD17-75-C-5056. AD-A142144

235. *van der Schalie W H* (1983) The acute and chronic toxicity of 3,5-dinitroaniline, 1,3-dinitroben-zene, and 1,3,5-trinitrobenzene to freshwater aquatic organisms. Techn report 8305, US Army Med Bioeng Res Develop Lab, Fort Detrick, Frederick, MD. AD-A138408

236. *Sprague J B* (1969) Water Res 3: 793

237. *Bailey H C, Spanggord R J, Javitz H S, Liu D H W* (1984) Toxicity of TNT wastewaters to aquatic organisms, vol IV. Chronic toxicity of 2,4-dinitrotoluene and condensate water. Final report prepared for US Army Med Res Develop Com by SRI Int, Menlo Park, CA. Contr No DAMD17-75-C-5056. AD-A153536

238. *Bailey H C, Spanggord R J, Javitz H S, Liu D H W* (1985) Toxicity of TNT wastewaters to aquatic organisms, vol III. Chronic toxicity of LAP wastewater and 2,4,6-trinitrotoluene. Final report prepared for US Army Med Res Develop Com by SRI Int, Menlo Park, CA. Contr No DAMD17-75-C-5056. AD-A164282

239. *Bringmann G, Kuhn R* (1980) Water Res 14: 231

240. *Smock L A, Stoneburner D L, Clark J R* (1976) Water Res 10: 537

241. *Schott C D, Worthley E G* (1974) The toxicity of TNT and related wastes to an aquatic flowering plant, *Lemna perpusilla,* Torr. Techn report 74016, Edgewood Arsenal, Aberdeen Prov Ground, MD. AD-778158

242. *Bentley R E, Dean J W, Ells S J, Hollister T A, LeBlanc G A, Sauter S, Sleight BH III* (1977) Laboratory evaluation of the toxicity of RDX to aquatic organisms. Final report prepared for US Army Med Res Develop Com by EG&G Bionomics, Wareham, MA. Contr No DAMD17-74-C-4101, AD-A061730

243. *Sullivan J H Jr, Putnam H D, Keirn M A, Pruitt B C Jr, Nichols J C, McClave J T* (1979) A summary and evaluation of aquatic environmental data in relation to establishing water quality criteria for munitions-unique compounds. Part 4: RDX and HMX. Final report

prepared for US Army Med Res Develop Com by Water and Air Res, Inc, Gainsville, FL. Contr No DAMD17-77-C-7027. AD-A087683

244. *van der Schalie W H* (1985) The toxicity of nitroguanidine and photolyzed nitroguanidine to freshwater aquatic organisms. Techn Report 8404, US Army Med Bioeng Res Develop Lab, Fort Detrick, Frederick, MD. AD-A153045

245. *Fisher D J, Burton D T, Paulson R L* (1989) Environ Toxicol Chem 8:

246. *Goodfellow W L, Burton D T, Graves W C, Hall L W, Cooper K R* (1983) Water Res Bull 19: 641

247. *Goodfellow W L, Burton D T, Cooper K R* (1983) Chemosphere 12: 1259

248. *Burton D T, Cooper K R, Goodfellow W L, Rosenblatt D H* (1984) Arch Environ Contam Toxicol 13: 653

249. *Stephan C E, Mount D I, Hansen D J, Gentile J H, Chapman G A, Brungs W A* (1985) Guidelines for deriving numerical national water quality criteria for the protection of aquatic organisms and their uses. Final report, PB85-227049. US Env Prot Agency, Office of Res and Develop, Wash, DC

250. *Gordon L, Hartley W R* (1989) Health advisory on 2,4,6-trinitrotoluene. Office of Drinking Water, US Env Prot Agency, Washington, DC

251. American Conference of Governmental Industrial Hygienists (1988) Threshold limit values and biological exposure indices for 1988-1989. ACGIH, 6500 Glenway Ave, Bldg D-7, Cincinnati, OH

252. Bureau of National Affairs, Inc (1988) Occupational safety and health reporter current report, vol 18, No 3, Wash, DC, p 233

253. Office of Emergency of Remedial Response, USEPA (1986) Superfund public health evaluation manual, EPA 540/1-86/060, OSWER Directive 9285. 4-1

Subject Index

The Handbook of
Environmental Chemistry

Edited by O. Hutzinger

Volume 5

Water Pollution

Part A

1990. Approx. 290 pp. 27 figs. 29 tabs.
Hardcover DM 198,– ISBN 3-540-51599-2

Contents: *G. F. Craun, Cincinnati, OH:* Epidemiologic
Studies of Organic Micropollutants in Drinking
Water. – *B. Allard, Linköping, Sweden;*
M. Falkenmark, Stockholm, Sweden: Water Quality
Genesis and Disturbances of Natural Freshwaters. –
H. L. Golterman, Arles, France;
N. T. de Oude, Strombeek-Bever,
Belgium: Eutrophication of Lakes,
Rivers, and Coastal Seas. –
W. T. Piver, Research Triangle Park,
NC; T. Lindstrom, L. Boersma,
Corvallis, OR: Mathematical
Models for Describing Transport
in the Unsaturated Zone of Soils.

Springer-Verlag
Berlin
Heidelberg
New York
London
Paris
Tokyo
Hong Kong
Barcelona

Environmental Toxin Series

Edited by S. Safe, O. Hutzinger

Volume 3

S. Safe, Texas A&M University, College Station, TX;
O. Hutzinger, University of Bayreuth;
T. A. Hill, Washington, DC (Eds.)

Polychlorinated Dibenzodioxins and -furans (PCDDs/PCDFs)

Sources and Environmental Impact, Epidemiology, Mechanisms of Action, Health Risks

1990. Approx. 160 pp. 8 figs. 32 tabs. Hardcover DM 174,–
ISBN 3-540-15552-X

Polychlorinated Dibenzodioxins and
-furans (PCDDs and PCDFs) are potent
environmental toxins. Environmental
exposures to these compounds in part-
per-billion (ppb) and part-per-trillion
(ppt) concentrations are of particular
interest. These exposures may arise from
bleached paper products, paper production process sludge, effluent waste water
from paper plants, consumption of food
and water from contaminated sources
and contact to paper products.

Springer-Verlag
Berlin
Heidelberg
New York
London
Paris
Tokyo
Hong Kong
Barcelona